	DATE DUE		

Hydrogen Power

Stuart A. Kallen

Energy and the Environment

ReferencePoint Press®

San Diego, CA

© 2010 ReferencePoint Press, Inc.

For more information, contact:
ReferencePoint Press, Inc.
PO Box 27779
San Diego, CA 92198
www. ReferencePointPress.com

Picture credits:
Cover: istockphoto.com
AP Images: 13
iStockphoto.com: 16, 76 (background)
Steve Zmina: 31–33, 45–46, 58–61, 75–78

LIBRARY OF CONGRESS CATALOGING-IN-PUBLICATION DATA

Kallen, Stuart A., 1955–
  Hydrogen power / by Stuart A. Kallen.
    p. cm. — (Compact research series)
  Includes bibliographical references and index.
  ISBN-13: 978-1-60152-073-9 (hardback)
  ISBN-10: 1-60152-073-5 (hardback)
  1. Hydrogen as fuel. I. Title.
TP359.H8K35 2009
665.8'1—dc22
                                                 2008049343

Contents

Foreword

❝ **Where is the knowledge we have lost in information?** ❞

—T.S. Eliot, "The Rock."

As modern civilization continues to evolve, its ability to create, store, distribute, and access information expands exponentially. The explosion of information from all media continues to increase at a phenomenal rate. By 2020 some experts predict the worldwide information base will double every 73 days. While access to diverse sources of information and perspectives is paramount to any democratic society, information alone cannot help people gain knowledge and understanding. Information must be organized and presented clearly and succinctly in order to be understood. The challenge in the digital age becomes not the creation of information, but how best to sort, organize, enhance, and present information.

ReferencePoint Press developed the *Compact Research* series with this challenge of the information age in mind. More than any other subject area today, researching current issues can yield vast, diverse, and unqualified information that can be intimidating and overwhelming for even the most advanced and motivated researcher. The *Compact Research* series offers a compact, relevant, intelligent, and conveniently organized collection of information covering a variety of current topics ranging from illegal immigration and deforestation to diseases such as anorexia and meningitis.

The series focuses on three types of information: objective single-author narratives, opinion-based primary source quotations, and facts and statistics. The clearly written objective narratives provide context and reliable background information. Primary source quotes are carefully selected and cited, exposing the reader to differing points of view. And facts and statistics sections aid the reader in evaluating perspectives. Presenting these key types of information creates a richer, more balanced learning experience.

For better understanding and convenience, the series enhances information by organizing it into narrower topics and adding design features that make it easy for a reader to identify desired content. For example, in *Compact Research: Illegal Immigration*, a chapter covering the economic impact of illegal immigration has an objective narrative explaining the various ways the economy is impacted, a balanced section of numerous primary source quotes on the topic, followed by facts and full-color illustrations to encourage evaluation of contrasting perspectives.

The ancient Roman philosopher Lucius Annaeus Seneca wrote, "It is quality rather than quantity that matters." More than just a collection of content, the *Compact Research* series is simply committed to creating, finding, organizing, and presenting the most relevant and appropriate amount of information on a current topic in a user-friendly style that invites, intrigues, and fosters understanding.

Hydrogen Power at a Glance

Hydrogen Power and the Obama Energy Agenda

In February 2009, President Barack Obama signed the American Recovery and Reinvestment Act, a $787 billion economic stimulus bill that includes $54 million in tax incentives for alternative fuel pumps including those that dispense hydrogen. Additionally, the Obama administration's "New Energy for America" plan sets out long-term energy goals such as ensuring that 25 percent of America's electricity comes from renewable sources by 2025 and investing $150 billion over 10 years to stimulate private clean energy projects. The how, what, and when of such spending has prompted considerable debate.

Hydrogen Production

Hydrogen molecules make up 75 percent of all matter on Earth, but hydrogen does not exist by itself in nature. It must be produced from fossil fuels or through an electrical process called electrolysis.

Fuel Cells

Fuel cells are high-tech batteries that convert the chemical energy of hydrogen into electricity. The only by-products from the process are heat and distilled water.

Hydrogen Research

The federal Hydrogen Fuel Initiative set aside $1.2 billion for hydrogen research over five years. The money has gone to national laboratories, college research facilities, and private companies involved in hydrogen development.

Fuel Cell Vehicles

A fuel cell vehicle converts hydrogen into DC electricity which powers an electrical motor that turns the wheels of the car. All major auto companies are developing fuel cell vehicles that might be on the road within 10 years.

Hydrogen Electricity

Fuel cells are being used today to provide electricity to private homes and other buildings. Someday millions of privately owned fuel cells might replace large centralized power plants.

Hydrogen and Global Warming

When burned, hydrogen produces zero carbon dioxide, a gas that contributes to global warming. However, most hydrogen produced today is made with fossil fuels, which adds carbon dioxide to the atmosphere.

The Future of Hydrogen

Scientists are working on a method to produce hydrogen from green algae, or pond scum, a technique that does not rely on fossil fuels or nuclear power.

Overview

66 With a new national commitment . . . the first car driven by a child born today could be powered by hydrogen, and pollution-free. 99

—George W. Bush, forty-third president of the United States.

66 Despite much hype to the contrary, a hydrogen economy is a long way off. 99

—Joseph J. Romm, physicist and executive director of the Center for Energy and Climate Solutions.

In 1965 American astronauts spent a record 8 days in space. Their *Gemini V* command module was powered by hydrogen fuel cells. The cells successfully provided 2 essential elements for space travel—electricity and water. This allowed NASA to continue with plans to land on the moon 4 years later. Without the hydrogen fuel cells, the moon landing would never have taken place.

Few Americans in 1965 had ever heard of hydrogen fuel cells. At the time of the *Gemini V* mission, gasoline was about 25 cents a gallon, the environmental movement was in its infancy, and most scientists were unaware of global warming. Only a very few very advanced thinkers imagined using hydrogen fuel cells to power cars or produce household electricity. Most Americans, if they thought about it at all, simply assumed that fossil fuels would take care of their energy needs for centuries to come. A lot has changed since the 1960s, when American drivers burned less than 10 million barrels of oil per day. In 2008 the United States consumed about 21 million barrels of oil a day, and 60 percent of that oil came from foreign sources. People have also come to understand that fossil fuels are nonrenewable resources that pollute the air and contribute to global warming. In addition, supplies of oil and natural gas are expected to run low before the end of the twenty-first century. An oil shortage would have a severe impact on the United States, where more than 97 percent of all cars, trucks, SUVs, and other vehicles run on gasoline or diesel fuel.

Americans also use a great deal of coal, another fossil fuel. About 50 percent of all electricity in the United States is generated by coal-burning power plants. Coal is cheap and abundant, and power plants in the United States use a billion tons of it every year. But coal is extremely harmful to the environment. When burned it releases a high concentration of carbon dioxide (CO_2) into the atmosphere. This gas is one of the major causes of global warming.

Advantages of Hydrogen

Researchers, scientists, and industry leaders throughout the world are predicting dire consequences if society does not find an alternative to oil and coal. Some believe that hydrogen, either burned as fuel or used in fuel cells, is the solution.

Hydrogen, or H_2, is an odorless, invisible gas and the most basic and abundant chemical element in the universe. It is most often found in water (H_2O), which consists of two hydrogen molecules (H_2) and one oxygen molecule (O). Although it is a major

> "
> In 2008, the United States consumed about 21 million barrels of oil a day, and 60 percent of that oil came from foreign sources.
> "

component of water, hydrogen by itself is extremely flammable. When burned, it produces three times as much energy as a similar amount of natural gas. And natural gas, while the cleanest fossil fuel, emits carbon dioxide when burned. By contrast, when hydrogen is burned it produces only heat and pure water.

As a flammable, nonpolluting gas, the advantages of hydrogen seem obvious. But there are many obstacles that must be overcome before hydrogen can satisfy society's increasing demands for electricity and transportation fuel.

How Is Hydrogen Produced?

Hydrogen does not exist by itself in nature. It cannot be dug up like coal or extracted through drilling like oil. Hydrogen needs to be obtained from other elements such as water or fossil fuels before it can be used as an energy source. There are two main methods used to produce hydrogen: One relies on natural gas, the other on electricity.

About 90 percent of the hydrogen made in the United States is derived from natural gas, which is a compound called a hydrocarbon. This compound consists of hydrogen and carbon molecules. In a complex process called steam reforming, the hydrogen is separated from the carbon. However, steam reforming is an inefficient way to produce hydrogen. It requires more energy from the natural gas than is available in the hydrogen that is produced. It would be cheaper and easier simply to burn the natural gas than use it to make hydrogen for fuel. In addition, natural gas is nonrenewable and contributes to global warming when burned.

Hydrogen can also be created from water in a technique called electrolysis, which utilizes electricity. This method involves a machine called an electrolyzer that uses direct current (DC) electricity to split the hydrogen and oxygen atoms from molecules of water.

> About 50 percent of all electricity in the United States is generated by coal-burning power plants.

While electrolysis is more energy efficient than steam reforming, the process still requires more energy than it produces. That makes electrolysis expensive. Hydrogen fuel produced by electric plants using fossil fuels or nuclear energy costs two to three times more than an equal amount of gasoline. Therefore, electrolysis that relies on electricity from traditional power plants is not the ideal way to produce hydrogen.

Environmentalists prefer harnessing renewable sources of energy for the electrolysis process. By using wind generators, solar panels, geothermal energy, or biomass, hydrogen can be produced while bypassing problems associated with nonrenewable fuels. However, these "green" energy sources only produce about 6 percent of the electricity in the United States and a tiny percentage of hydrogen.

Storage and Transport

Even if enough hydrogen could be produced through nonpolluting methods, there are technical problems concerning storage and transportation. As the lightest gas in the universe, hydrogen floats off into space unless it is tightly contained. While oil is easily stored in metal barrels or plastic containers, hydrogen must be compressed into tanks much like those that contain helium used to inflate balloons.

Leakproof storage tanks for hydrogen are expensive, and compressing the gas to fit in the tanks requires a great deal of natural gas. Without compression, however, hydrogen takes up a lot of space, as chemistry professor Lev Gelb explains: "If you had a kilogram of hydrogen . . . you'd have to store it in about 100 big balloons, if you can picture that. A kilogram of gasoline, on the other hand—

> " **Hydrogen is an odorless, invisible gas and the most basic and abundant chemical element in the universe.** "

that would [fit into] a small container."[1] Hydrogen can also be stored by cooling it to -423°F (-253°C). This, too, requires a substantial quantity of natural gas. Discussing the problems of hydrogen production and storage, Gelb states: "The case has been made persuasively that you'd be better off just burning the natural gas, rather than going to the trouble of producing hydrogen from natural gas and going through all the problems associated with its storage and transport."[2]

Can Hydrogen Vehicles Reduce Dependence on Fossil Fuels?

Those searching for answers to hydrogen problems received a boost in 2003 when President George W. Bush implemented the Hydrogen Fuel Initiative. This program dedicated $1.2 billion over 5 years to develop hydrogen production, delivery, and storage techniques, and hydrogen fuel cell technologies. The aim of the program is to develop hydrogen-powered cars with a 300-mile (483km) driving range.

Hydrogen is used in two ways to power vehicles. Combustion vehicles burn hydrogen in modified internal combustion engines. An internal combustion engine can burn hydrogen with a few minor modifications, much like natural gas vehicles do today. Such methods are utilized by the BMW Hydrogen 7, a luxury automobile built in Germany.

A second type of hydrogen car, called a fuel cell vehicle (FCV), uses what is called a PEM fuel cell, which stands for polymer electrolyte membrane or proton exchange membrane. The PEM is one of six types of fuel cells, but it holds the most promise for powering vehicles and homes. The PEM fuel cell converts hydrogen fuel into DC electricity, which is used to power an electric motor that turns the car's wheels. PEM fuel cells are used in the GM Sequel and in the Honda FCX Clarity.

Ford Motor Company has invested over $1 billion in FCVs. Its former chair, William Clay Ford, is convinced that Ford's new technology will help end the dependence on oil, stating, "I believe fuel cells will finally end the 100-year reign of the internal combustion engine."[3]

> " Hydrogen fuel produced by electric plants using fossil fuels or nuclear energy costs two to three times more than an equal amount of gasoline. "

Other Fuel Cell Vehicles

While automakers have invested massive sums in hydrogen-powered automobiles, only about 200 such vehicles were on the road in the United States in 2007. However, fuel cells can be used in many types of vehicles, and hydrogen city buses are being built in the United States, Germany, and Japan.

In 2007 the American company ISE, a leading supplier of hybrid buses and trucks, sold 10 hydrogen fuel cell buses to the city of London. Five are powered by hydrogen combustion engines and 5 run on fuel cells. Tests will be conducted to determine which type of hydrogen vehicle performs better in city traffic, and there are plans in London to add up to 60 more FCV buses in the near future. Commenting on the FCV bus program, called the London Hydrogen Partnership, Mayor Ken Livingstone said:

> Hydrogen fuel cells could offer a real alternative to diesel [powered buses] in the future. The high cost of the vehicles is the major barrier at the moment but the greater

The city of Hartford, Connecticut, uses a few hydrogen-powered buses. All of the major auto companies are developing fuel cell vehicles that might be on the road within 10 years.

the demand for vehicles, the more the costs will come down. I would call on the manufacturers to gear up for this change, as hydrogen vehicles are a real and viable option for London.[4]

Fuel cells can also be used in much smaller modes of transportation. In 2007 the British company Intelligent Energy Ltd. began selling its ENV Bike, a motorcycle powered by a small fuel cell. The cycle operates silently, has a top speed of 50 miles per hour (80kph), and can be purchased for about $10,000.

In China scientists have developed the PHB, a hydrogen-powered bicycle. The bike comes with traditional pedals but can also be ridden at 15 miles per hour for 60 miles (24kph for 97km) with an assist from a fuel cell that produces about 200 watts of electrical power. The bike now costs around $3,000, but the Chinese hope to mass-produce the PHB and sell it for about $500 by 2012.

More hydrogen vehicles are being developed every year. However, it is estimated that it could take 40 years or more for hydrogen to replace fossil fuels. Hydrogen skeptics point out that there are many other technologies, such as electric vehicles and plug-in gas-electric hybrids, that show much more promise in reducing oil consumption. And these vehicles use fuels that are readily available at a relatively low cost. In addition, tens of thousands of hydrogen fueling stations would need to be built or installed at traditional gas stations, which would cost billions of dollars. Discussing the problems with hydrogen-powered vehicles, energy expert Steve Plotkin states, "The transition to a new fuel involves a huge change in the energy system. It will never work in the real world unless costs come way down."[5]

> " An internal combustion engine can burn hydrogen with a few minor modifications. "

Can Hydrogen Supply the World's Electricity Needs?

The same type of fuel cell used to power cars can also be put to work providing electricity for homes, offices, and other buildings. In Japan, a nation with no oil or natural gas reserves, fuel cells are already being used in this manner. In 2008 Nippon Oil sold about 10,000 fuel cells for private use. These mini–power plants, each about the size of a typical central air conditioner, are the world's smallest home electricity systems. Although they are expensive, more than $19,000 each, the company plans to make 150,000 fuel cells by 2015. This is expected to lower the price to around $5,000 due to the economic benefits of mass production. Whatever their cost, the fuel cells save an average Japanese family about $650 annually on their electric bills and so eventually pay for themselves.

How Will Hydrogen Power Impact Global Warming?

Demand for electricity is growing in countries all over the world, causing carbon dioxide emissions to rise. In the United States, where one-fifth of all carbon dioxide comes from cars and trucks, CO_2 emissions are expected to grow 15 percent by 2035. More carbon dioxide will add to the long list of problems already blamed on global warming. These include the thinning of the polar ice cap, rising sea levels, and the shrinking of glaciers around the world. Global warming is also expected to contribute to more severe weather. This includes increasingly severe droughts, wildfires, heat waves, and hurricanes, along with the spread of tropical diseases to northern climates.

To deal with these serious problems, governments and industries throughout the world have pledged to lower their CO_2 emissions in the coming years. There is little doubt that a widespread use of hydrogen vehicles and fuel cell power plants would slow the effects of global warming. But the hydrogen has to be produced in carbon-neutral ways, that is, methods that do not emit carbon dioxide. Therefore, the widely used processes utilizing natural gas or coal-based electricity would have to be discontinued. Discussing this problem, Geoffrey B. Holland, cofounder of the group Hydrogen 2000, and James J. Provenzano, president of Clean Air Now, state: "[Hydrogen] *is everywhere*, but not in the 'free' form that makes it useful as a nonpolluting fuel. What the world needs is carbonless, free hydrogen."[6]

Renewable Power

There are ways to produce free hydrogen wherever there is renewable electricity and water. One such place is Prince Edward Island in Canada, where windmills are being used for hydrogen electrolysis. At the Wind-Hydrogen Village Project, the steady wind blowing from the Gulf of St. Lawrence is continually creating hydrogen that is being used in the island's hydrogen combustion vehicles and fuel cell buses.

In the United States the central part of the country that stretches from Texas to North Dakota is known as a wind corridor, where turbines are already producing abundant and inexhaustible power. Speaking of this region, Holland and Provenzano write, "Wind's enormous potential alone puts the goal of a hydrogen economy powered by clean, renewable energy within reach,"[7] with a corresponding reduction in global warming gases.

Hydrogen production from solar energy is also showing great promise. Solar photovoltaic cells (PVs) produce electricity from the sun. Worldwide

> " The same type of fuel cell used to power cars can also be put to work providing electricity for homes, offices, and other buildings. "

This house uses solar power for some of its energy needs. By using solar panels to produce solar power, hydrogen can be produced while bypassing problems associated with nonrenewable fuels.

production of solar cells has increased an average of 60 percent annually since 2001, and demand for PVs remains strong. Today hundreds of people throughout the world are already using their solar cells to create hydrogen through electrolysis, and the number is growing every year. Plans for solar hydrogen fuel cell systems are available on the Internet, and videos of solar hydrogen–powered homes may be seen on YouTube.

Hydrogen can also be produced through other renewable sources, including biomass and hydropower. But if nothing is done to move society toward free hydrogen production, the United Nations' Intergovernmental Panel on Climate Change (IPCC) warns that CO_2 will rise by an additional 90 percent worldwide by 2030. Former president George W. Bush believed hydrogen could reverse that trend. In 2003 he said, "Hydrogen power will dramatically reduce greenhouse gas emissions, helping this nation take the lead when it comes to tackling the long-term challenges of global climate change."[8]

What Is the Future of Hydrogen Power?

Despite the former president's optimism, science must overcome several complicated problems before hydrogen can reduce global warming and provide an alternative to fossil fuels. On the production end, there are simply not enough windmills or solar panels available to produce hydrogen for the 300 million automobiles expected to be on the road in the United States in 15 years. The most optimistic predictions say that wind will supply 20 percent of the power in the United States by 2030 and solar could supply about 10 percent. These figures do not take into account the huge upsurge in demand for electricity that is expected to occur when millions of people switch to electric vehicles, plug-in hybrids, and FCVs. But those who promote hydrogen believe that technology will provide answers to these problems in the future, as Holland and Provenzano state: "[Solar energy] is a technical arena that is pushing the envelope in many different directions. Fifty years from now, the world may be harvesting sunlight in ways that haven't even been imagined yet."[9]

> **Today, hundreds of people throughout the world are already using their solar cells to create hydrogen.**

The "harvested sunlight" could be used to make hydrogen. Some envision a time when nearly every home has solar panels and mini-windmills that produce hydrogen when the sun is shining and the wind blowing. They hydrogen will be stored in tanks and used to power fuels cells for cars and home electrical needs.

Home fuel cells like those being sold in Japan might someday change the way electricity is used throughout the world. The small units could be used for electrical power in remote regions of the Earth where large power plants and miles of transmission lines do not exist. In urban areas, the unused power from the fuel cells could be fed into the interconnected network of high voltage wires known as the electrical grid. Utility companies, instead of building large, expensive, centralized power plants, could direct the fuel cell power into transmission lines and transport it to areas where it is needed.

> There are simply not enough windmills or solar panels available to produce hydrogen for the 300 million automobiles expected to be on the road in the United States in 15 years.

On the automotive front, hydrogen fueling stations may be more than a distant dream. According to the Hydro Cars & Vehicles Web site, a station to fuel hydrogen vehicles costs between $400,000 and $2 million to build. Larry Burns, vice president of General Motors Research and Development and Strategic Planning, believes that "a network of 12,000 hydrogen stations in the United States would put 70 percent of the U.S. population within 2 miles of a fueling station."[10] If the stations cost $2 million each, the total would come to $24 billion. To put that large dollar amount in perspective, the budget for U.S. Department of Defense in 2008 was $657 billion. This means that 12,000 hydrogen fuel stations could be built for what the Pentagon spends in about 2 weeks.

China, Japan, Korea, and several European nations plan to build a hydrogen fueling infrastructure by 2012. While interest remains high in electric vehicles and gas-electric hybrids, there is little doubt that hydrogen fuel cells will be powering cars and homes in the future.

Can Hydrogen Vehicles Reduce Dependence on Fossil Fuels?

> **Hydrogen-fueled super-efficient vehicles will be safer and cleaner, cost less to drive, [and] cost about the same to buy.**
>
> —Amory P. Lovins, chairman of the board, Hypercar, Inc.

> **There is little point in manufacturing a fleet of very expensive hydrogen-fueled vehicles if there is no place to buy hydrogen produced under environmentally-responsible conditions.**
>
> —Rex A. Ewing, author and renewable energy expert.

On August 15, 2007, the Ford Buckeye, powered by a hydrogen fuel cell, raced down the course at the Bonneville Salt Flats in Utah. When it hit a speed of 207.2 miles per hour (333.5kph), the Buckeye went into the record books as the fastest fuel cell vehicle in history. The record-breaking demonstration was a result of a collaboration called 999 Hydrogen Fusion, a project of students at the Ohio State University Center for Automotive Research and Ford Motor Company.

More than a public relations stunt, 999 Hydrogen Fusion proved that hydrogen fuel cells can compete with traditional gasoline-powered internal combustion engines and might someday replace them. As Ford project leader Matt Zuehlk stated, "All of the technology used on this vehicle . . . can be made to fit production cars down the road."[11]

Hydrogen-Powered Machinery

Hydrogen molecules make up 70 percent of the mass of planet Earth. Because the flammable gas is found in plants, animals, and water, it is known as the forever fuel because it can never run out. The ability of hydrogen to power vehicles was understood as early as 1820 by English inventor William Cecil. That year Cecil presented a paper to the Cambridge Philosophical Society entitled "On the Application of Hydrogen Gas to Produce Moving Power in Machinery."

> "Hydrogen fuel cells can compete with traditional gasoline-powered internal combustion engines and might someday replace them."

Cecil never built hydrogen-powered machinery, but in 1923 J.B.S. Haldane, another scientist from the United Kingdom, proposed using windmills to separate hydrogen from oxygen in water. Haldane wrote that the hydrogen gas could be liquefied and stored in huge underground tanks to power internal combustion engines. He understood that hydrogen was an extremely efficient fuel with 3 times more energy per pound than petroleum. Although few at the time believed that petroleum and other fossil fuels would ever run out, Haldane believed a wind-hydrogen system would be an important alternative source of power in the future.

In 1965 Haldane's ideas inspired a 16-year-old inventor named Roger Billings to modify his father's old Model A Ford to run on hydrogen. Twelve years later Billings drove his hydrogen-powered Cadillac in President Jimmy Carter's inauguration parade. Billings continued his work, creating factory machinery and dozens of cars that ran on hydrogen. His unceasing promotion of hydrogen cars earned him the name Dr. Hydrogen. But as Billings and other promoters understand, there are obstacles to overcome before hydrogen-powered cars can replace ones with gasoline-powered engines.

Bumps in the Road

Worldwide there are more than 700 million vehicles, with about 66 million new ones produced every year. These vehicles consume more than 600 billion gallons (2.27 trillion L) of gasoline and diesel fuel annually. About one-third of that fuel is burned in the United States, where there were over 250 million registered vehicles in 2006. These vehicles were produced by an international auto industry that is the biggest and most diverse manufacturing entity in the history of the world. Getting an industry of this size to make major changes in the way vehicles are designed, built, and powered is a staggering task.

In the United States the "Big Three" automakers—Ford, General Motors, and Chrysler—showed little interest in hydrogen vehicles when profits were high throughout the 1990s and early 2000s. But by 2008, the Big Three were losing billions as a result of high gas prices and a product lineup heavy with inefficient trucks and SUVs. Discussing the predicaments faced by the Big Three, Holland and Provenzano write:

> Virtually every auto company . . . recognized the need to wean itself from oil. The vision most of them share revolves around hydrogen. Ethanol, biodiesel, and electricity will all have their place as a source of automotive power, but . . . the fuel that will prevail will be [hydrogen], the one that is storable in large volume for use on demand, is virtually limitless in supply, and is essentially pollution free.[12]

But hydrogen cars face several obstacles, the first of which is cost. Fuel cells use large amounts of platinum, a nonrenewable precious metal that costs around $2,000 an ounce. This drove the cost of a 200-horsepower fuel cell to about $75,000 in 2008. The fuel cell alone, big enough to power a midsize 4-door sedan, costs about 3 times as much as a Toyota Prius hybrid.

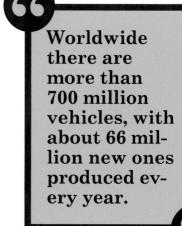

"Worldwide there are more than 700 million vehicles, with about 66 million new ones produced every year.

Overcoming Barriers

Most researchers are sure that there are ways to bring down the cost of fuel cells. However, there are practical matters that are preventing widespread sale of fuel cell vehicles. One of the immediate problems is not hard to understand but has proved difficult to overcome. Fuel cells produce water as a by-product of operation. When the temperature is below 32°F (0°C), the water freezes and prevents the fuel cell from starting. Heating devices can be applied to the fuel cell but in places like the Upper Midwest where winter temperatures dip to -30°F (-34°C), such warming methods consume large amounts of energy.

Once the technical barriers are overcome Taiyo Kawai, general manager of the Fuel Cell System Engineering Division for Toyota, believes that other measures must be accomplished before fuel cell vehicles are ready for mass production: "First, there is the issue of vehicle marketability. . . . [The] cost of the vehicle must be at a level that the market can support. As with other vehicles today fuel cell vehicles must be packaged in a way that is attractive to the consumer. . . . Finally, there must be a consumer shift to understand and prefer environmental vehicles."[13]

Hydrogen Cars on the Road

In order to overcome these problems, the Big Three as well as Honda, Hyundai, Nissan, Toyota, and Volkswagen are building fuel cell vehicles. Some companies are providing them to select members of the public for free or at a relatively low cost. This not only generates publicity for the auto maker but allows the car to be road tested in daily driving conditions. In 2008 there were more than 200 such vehicles in use, with about 175 in California, where warm temperatures are conductive to fuel cell operation.

One such vehicle is the Honda FCX Clarity, which achieves the equivalent of 68 miles per gallon (29km/L) of gas and produces zero emissions. Customers in Southern California pay about $600 a month for a 3-year lease, but this does not come close to covering the actual price of the car, which costs about $2 million. Discussing the astronomical price, Honda spokesman Todd Mittleman states, "Widespread commercialization is nothing we'll see in the next few years, but it is the ultimate solution."[14]

Hydrogen Internal Combustion Engines

BMW took a different approach to building a hydrogen car. With only a few, inexpensive modifications, the German automaker was able to modify its standard gasoline-powered internal combustion engines to burn hydrogen. This eliminated the need for expensive fuel cells that might have problems running in cold temperatures. The cars work so well that in 2004 the German automaker won nine world speed records with its hydrogen-powered, 12-cylinder BMW H_2O race car.

The same engine that powers the H_2O is now installed in the BMW Hydrogen 7. This 12-cylinder, 260 horsepower car runs on liquid hydrogen

(LH_2) rather than the hydrogen gas used by fuel cell vehicles. Commenting on the abilities of the vehicle, BMW board member. Klaus Draeger stated, "[Hydrogen] is the only energy carrier that can replace fuels like diesel or gasoline in the future. . . . The [hydrogen] internal combustion engine is the only way to combine the BMW typical agility with the BMW typical dynamics."[15]

Most standard car engines can be made to run on hydrogen. The conversion only requires a few modifications to the engine fuel injection system. But there are technical problems, because liquid hydrogen must be kept at a temperature of -423°F (-253°C), about the same frigid coldness as outer space. This requires the addition of large, super-insulated, high-pressure fuel storage tanks. These tanks, which have to be crash proof so they do not explode in an accident, add considerable cost to the BMW.

Beyond the world of fast, agile, luxury sedans, hydrogen is being used in the world of public transportation. Ford began building a fleet of 10-cylinder, hydrogen internal combustion engine shuttle buses in 2006. Eight of them are currently in use at the Orlando International Airport in Florida. Commenting on the future of hydrogen combustion engines, Paul Scott, chief scientist for a company that makes alternative fuel buses, states: "The hydrogen IC [internal combustion] is the closest thing you can get to zero pollution in a combustion engine. . . . In mass production, we estimate [our buses] will only cost about 20 percent more than a traditional transit coach. The hydrogen-IC engine is like having one foot set in the past and the other foot in the future."[16]

> " The Honda FCX Clarity . . . achieves the equivalent of 68 miles per gallon (29 km/L) of gas and produces zero emissions. "

Fueling Hydrogen Vehicles

BMW is testing about 20 Hydrogen 7s in the United States. Several have been given to big-name Hollywood stars, including Jay Leno, Ed Norton, Brad Pitt, and Cameron Diaz. However, the cars are not practical for everyday use. When a Hydrogen 7 runs low on fuel, the driver must turn it over to a trained BMW technician, who takes it to the nearest liquid hydrogen fueling station in Oxnard, California, about 60 miles (97km) from Hollywood. Due to traffic conditions, the round-trip often takes more than 4 hours. But Hydrogen 7 drivers need not worry about running out of fuel, since the BMW can switch effortlessly to run on gas. With this duel-fuel capacity, the Hydrogen 7 can travel 130 miles (209km) on LH_2 and 310 miles (500km) on gasoline.

> **Several [Hydrogen 7s] have been given to big-name Hollywood stars, including Jay Leno, Ed Norton, Brad Pitt, and Cameron Diaz.**

Big stars are also leasing the Clarity from Honda. These drivers can fill up at any of the 24 hydrogen refueling stations in Southern California. And there are plans to build 10 more such stations in the region. But Honda wants to move beyond centralized fueling stations. To help do so, the company developed an experimental Home Energy Station in Torrance, California. This residence uses natural gas to produce hydrogen, which is used to power a fuel cell vehicle and a small fuel cell electrical system for the home. The company is hoping to perfect a solar-powered system that will someday allow owners to fill up their cars inside their own garages with CO_2-free hydrogen.

Not the Way to Go?

There are those who remain skeptical about hydrogen's reducing dependence on fossil fuels. James Woolsey is the former head of the Central Intelligence Agency and chair of the Clean Fuels Foundation, a group that promotes electric and hybrid vehicles. Woolsey believes it is in the national security interests of the United States to stop global warming and reduce dependence on imported oil. But Woolsey believes the cost of establishing a hydrogen system is too expensive, as he stated in November 2007:

> Hydrogen and fuel cells are not the way to go. The decision by the Bush administration and the State of California to follow the hydrogen highway is the single worst decision in the past few years. . . . Someone may find a way to make hydrogen cheaply available but that is still some time away. Our estimates put building a hydrogen

infrastructure at one trillion dollars. In the meantime you have to bring down the cost of a fuel cell vehicle by a factor of about 40 or 50 to make it affordable. Joining the hydrogen highway for families is a poor idea. For large fleets with a single filling facility it might work.[17]

Others disagree with Woolsey's assessment, saying that his trillion-dollar estimate is based on the concept of building a nationwide hydrogen pipeline system similar to the one used to transport natural gas. But a study conducted jointly by Ford Motor Company and the U.S. Department of Energy points out that "hydrogen can be delivered to fuel-cell vehicles at lower costs by producing and installing . . . small-scale electrolyzers at the local fueling station or fleet operator's garage."[18] In this way filling stations could produce hydrogen and sell it along with gas and diesel fuel. And the electrolyzers could even be equipped with wind or solar technology to produce pollution-free hydrogen. In this way hydrogen could be introduced slowly, eventually making gas and diesel pumps obsolete. Larry Burns at General Motors feels this would be the best solution for reducing oil consumption:

> **Filling stations could produce hydrogen and sell it along with gas and diesel fuel.**

If you woke up tomorrow and all 220 million cars and trucks in the United States had been hybridized to the degree that the Prius has—all getting 25 percent better [gas] economy—in six years we would be consuming the same amount of petroleum that we are right now. Fuel cells create a better automobile that's 50 percent more energy-efficient overall and sustainable from energy and safety perspectives.[19]

For those who criticize hydrogen, Mike Moran, founder of the consulting firm Automotive Partnership, points out that oil shortages and global warming will be huge problems for the next generation and offers some advice: "I think that anything that moves forward mankind's understanding has got to be a good thing. All the nay-sayers should just shut up."[20]

Primary Source Quotes*

Can Hydrogen Vehicles Reduce Dependence on Fossil Fuels?

Bracketed quotes indicate conflicting positions.

* Editor's Note: While the definition of a primary source can be narrowly or broadly defined, for the purposes of Compact Research, a primary source consists of: 1) results of original research presented by an organization or researcher; 2) eyewitness accounts of events, personal experience, or work experience; 3) first-person editorials offering pundits' opinions; 4) government officials presenting political plans and/or policies; 5) representatives of organizations presenting testimony or policy.

"These days . . . gasoline is in question . . . electric power is being considered and now we have hydrogen in the mix. Which fuel will emerge victorious? Probably the one that . . . gets to the people fastest and cheapest."

—Jay Leno, "Will Hydrogen Fuel Our Future? Jay Spends 10 Days Behind the Wheel of BMW's Hydrogen 7," *Popular Mechanics,* January 2008.

Leno is a comedian, auto enthusiast, and author of the column Jay Leno's Garage in *Popular Mechanics.*

"Hydrogen would cost three or four times more than gasoline. Fuel cells cost at least five times as much as standard combustion engines and aren't nearly as durable. It's also proving tough to build hydrogen storage systems."

—Anne Broache, "Feds Propose $100 Million Hydrogen Prize," CNET, April 27, 2006. http://news. cnet. com.

Broache is a staff writer for CNET news.

"Hydrogen is no longer simply a glorious vision—it is happening now, in cities around the world—including here in the United States. . . . [The] massive roll-out of FCVs [fuel cell vehicles] by 2020 is absolutely achievable."

—Duncan Macleod, "From Research to Reality," Shell.com, June 3, 2008. www.shell.com.

Macleod is global vice president of Shell Hydrogen.

66 [The] fantasy of cost-effective hydrogen fuel cell vehicles is just a distraction from the real work that needs to be done: perfecting electric and hybrid natural gas/electric vehicles, charged . . . by clean and renewable . . . power. **99**

—Charlie White, "SHIFT: Hydrogen Fuel Cell Vehicles Are a Fraud," DVICE, July 31, 2008. http://dvice.com.

White is deputy editor of the technology and consumer electronics Web site DVICE.

66 For the first time in human history, we have within our grasp a ubiquitous form of energy . . . [hydrogen] the forever fuel. **99**

—Jeremy Rifkin, *The Hydrogen Economy*. New York: Jeremy P. Tarcher/Putnam, 2002.

Rifkin is a best-selling author and an authority on cutting-edge technology.

66 Long after fossil fuels run out, hydrogen will remain. Taken from the world's rivers and oceans, it will keep our wheels turning when imported oil and coal-fired boilers are ancient history. **99**

—Michael A. Peavey, *Fuel from Water*. Louisville: Merit, 2003.

Peavey is a science writer whose fields of expertise include hybrid electric cars and fuel cell technology.

66 The outrageously high costs of fuel-cell cars and hydrogen fuel, combined with the non-existence of a hydrogen distribution and sales infrastructure . . . make the possibility of mass consumer purchases of hydrogen fuel-cell vehicles a non-starter. 99

—Robert Zubrin, "The Hydrogen Hoax," *New Atlantis,* Winter 2007. www.thenewatlantis.com.

Zubrin is an aerospace engineer and the president of Pioneer Astronautics, a research and development firm.

66 Fuel cell–powered vehicles have the potential to revolutionize power and mobility. Not only would this create very many, technologically advanced new jobs, but this diversity would also dramatically stabilize the economy. 99

—Chris Borroni-Bird, "Hydrogen Power: A Discussion with Chris Borroni-Bird," Scientific American Frontiers, PBS. www.pbs.org.

Borroni-Bird designs concept cars for General Motors and is a leading expert on fuel cell vehicles.

Can Hydrogen Vehicles Reduce Dependence on Fossil Fuels?

- Burning fossil fuels adds sulfur oxides, nitrogen oxides, soot, and other pollutants to the air. Hydrogen is **nontoxic**, so it is not harmful to breathe.

- Fuel cell vehicles are **50 percent** more energy efficient than gas-powered cars.

- The cheapest way to produce hydrogen is by using electricity from **wind turbines**.

- A 200-horsepower fuel cell cost about **$75,000** to produce in 2008.

- A high-pressure fuel tank used in a hydrogen vehicle costs up to **$30,000** and weighs as much as a small car, about **2,680 pounds** (1,216kg).

- Converting every vehicle in the United States to hydrogen would require so much electricity that the country would need enough wind turbines to cover half the state of California or as many as **1,000 extra nuclear power stations**.

- Hydrogen fuel injection systems bolt onto diesel engines and inject a small amount of hydrogen into cylinders, improving fuel economy by **10 to 30 percent**.

How a Fuel Cell Works

The PEM, or proton-exchange membrane fuel cell is typical of those used in fuel cell vehicles. Protons pass through the proton-exchange membrane while the electrons are forced to go around it. When hydrogen and oxygen are forced together in the cell, electricity is produced. By-products of the process include heat and water.

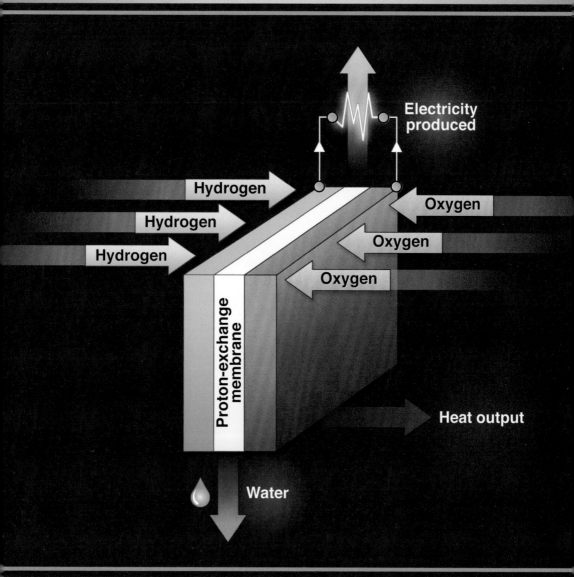

Source: Center for Fuel Cells, "Fuel Cell Technology," 2008. www.che.sc.edu.

Limited Hydrogen Fueling Stations in the United States

Only a few states have hydrogen fueling stations and all except California have fewer than 10. This is a major problem for consumers wishing to buy hydrogen-powered vehicles.

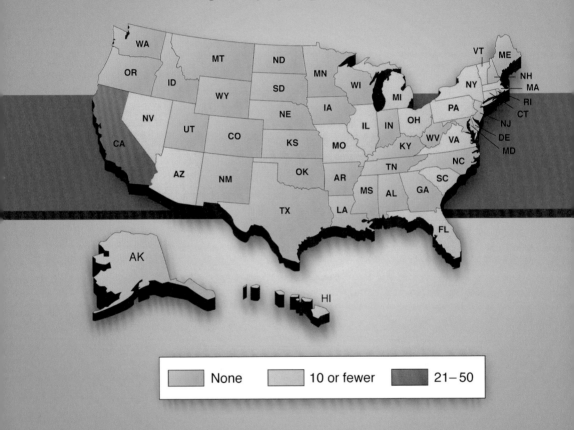

| None | 10 or fewer | 21–50 |

Source: Alternative Fuels & Advanced Vehicle Data Center, "Hydrogen Fueling Station Locations," September 18, 2007. www.afdc.energy.gov.

- It takes **four times more** electricity to make hydrogen to fill up a fuel cell vehicle than it takes to charge a battery power car.

- Honda spent **$400 million** to develop and build only 200 FCX Clarity fuel cell vehicles.

How a Fuel Cell Vehicle Works

This illustration shows a typical fuel cell vehicle design. Hydrogen is stored in a hydrogen tank (1); this provides fuel for a fuel cell (2) which send electricity to a lithium-ion battery (3). The battery provides power to the motor (4), which regulates the amount of power sent to the electric drive motor (5) which sends power to the car's wheels and propels the car down the road.

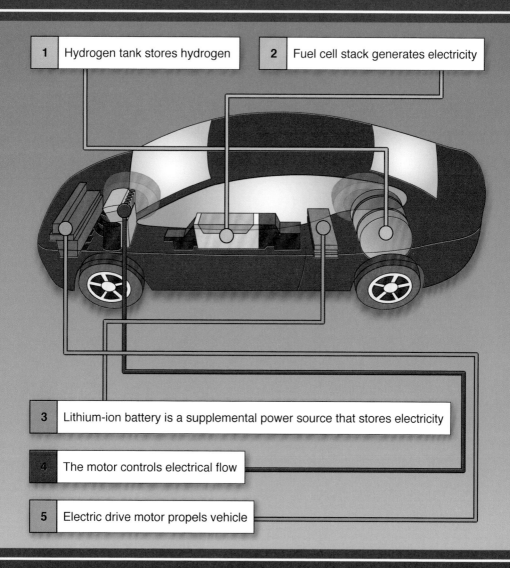

1   Hydrogen tank stores hydrogen

2   Fuel cell stack generates electricity

3   Lithium-ion battery is a supplemental power source that stores electricity

4   The motor controls electrical flow

5   Electric drive motor propels vehicle

Source: Honda, "How FCX Clarity Works," 2008. http://automobiles.honda.com.

Can Hydrogen Supply the World's Electricity Needs?

❝ As long as you have hydrogen to run through these fuel cells it is very cheap and efficient. And you can store it in a propane tank in a back yard. . . . This will be very common in the next few years.❞

—Robert Plarr, builder and designer of a hydrogen-powered house in Taos, New Mexico.

❝ Electricity from hydrogen fuel cells will cost about four times as much as electricity from the grid.❞

—Ulf Bossel, founder of the European Fuel Cell Forum.

In 1963, Australian electrochemist John O'M. Bockris developed a concept called the "hydrogen economy." In his vision hydrogen would be part of a system of harmonious, balanced elements providing electricity to the industrial world. Bockris proposed building a web of nuclear power plants and solar collectors on platforms in the ocean or in very remote, uninhabited regions of North America, Canada, and central Australia. The electricity generated at these sites would be converted to hydrogen, which would be piped to population centers. As Bockris writes: "[With the hydrogen economy] it could be cheaper to convert electrical energy, which will be a product of solar and atomic reactors, to hydrogen at the energy source. Thereafter, the hydrogen would be transmitted through pipes . . . and converted back to electricity at the site of use [with fuel cells]."[21]

Bockris worked as a consultant for General Motors when he first expressed his futuristic vision. He was disregarded at the time, but more than 45 years later, General Motors—and hundreds of other companies—are creating their own versions of the hydrogen economy. But in this usage the term *economy* has nothing to do with money or other economic issues. Instead, the hydrogen economy denotes an integrated system of components, including wind turbines (windmills), solar collectors, electrolyzers, and fuel cells that work together to create energy.

Power Grid at Risk

Several critical forces are driving the move to a global hydrogen economy, the first being energy security. Industrial societies rely heavily on coal, natural gas, oil, and nuclear power to produce electricity. The power plants that use these fuels are large and centrally located. The power they produce is distributed through countless miles of power lines. The power plants, transmission lines, and the computer systems that control the power grid are tempting targets for terrorists. Experts say there are also threats to the natural gas pipelines that provide the basic energy for about a quarter of the power plants in the United States. After the September 11, 2001, attacks the Department of Energy noted the threat, issuing a report that states, "A relatively small group of dedicated, knowledgeable individuals could bring down [the power grid] in almost any section of the country."[22] Such an attack could devastate major cities, cause thousands of deaths, and cost tens of billions of dollars in damages.

Reliance on fossil fuels creates other vulnerabilities as well. Countries such as Japan and Korea have to import nearly all of the fuels they use to produce electricity. European countries are heavily reliant on natural gas produced in Russia. Because of this, Russia can dictate the price of natural gas and cut off supplies if that action suits its purposes. In the winter of 2006, for instance, Russia cut off gas supplies to Ukraine and Moldova and threatened to cut off gas supplies to Belarus, Georgia, Lithuania, and parts of Germany during price negotiations.

> " The hydrogen economy denotes an integrated system of components, including wind turbines (windmills), solar collectors, electrolyzers, and fuel cells that work together to create energy. "

The United States produces much of its own natural gas. But in 2005 oil giant Exxon's former chief executive Lee Raymond stated that supplies will not meet the growing natural gas demand in the coming years. And officials at Dutch Shell say worldwide natural gas shortages could occur by 2025. Therefore, energy security is dependent on finding alternative sources to natural gas and other fossil fuels used for electricity production.

> **Power plants, transmission lines, and the computer systems that control the power grid are tempting targets for terrorists.**

Even without fuel shortages and terrorist attacks, much of the power grid in the United States is aging and inefficient. Half the power plants in the country are decades old and burn polluting coal, which contributes to global warming. Storms, wildfires, and equipment breakdowns have resulted in massive power failures and brownouts in recent years. The economic damages from these power outages cost hundreds of millions of dollars. Even as the power grid deteriorates, consumer demand for electricity in the United States is expected to grow 45 percent by 2030. And current power plants and transmission lines are already operating near full capacity in many regions.

Micro–Power Plants

A hydrogen economy based on renewable energy generation could solve many of the problems associated with the power grid. Hydrogen could be produced by solar panels, wind turbines, hydroelectric dams, or through biomass, a type of energy produced by agricultural waste. Power generation would be decentralized, since each individual consumer would operate a fuel cell power plant on his or her own property. The cell could be either permanently attached to the house or part of a fuel cell vehicle. Terrorist threats to the power grid would be lessened, global warming would be reduced, and there would be less reliance on imported fossil fuels.

If millions of people owned fuel cell power plants, they could participate in a system called distributed generation. This innovative new way to deliver electricity is the opposite of the system in place today. Instead of power being generated at a centrally located plant and delivered through power lines, fuel cell power plants would be located on or near the site where power is used, including offices, factories, shopping malls, or homes and apartments. Because fuel cells come in modules, power users could customize the micro–power plants to fit their needs. If more power was needed, extra modules could be purchased and added to the fuel cell. Those who had excess power could upload it onto the power grid, as some people do with solar panels today. This could even result in power companies paying the fuel cell owner for the extra electricity, which would be sold to people without fuel cells.

A Hydrogen-Powered House

The concept of decentralized power generation is already in practice in some places. In 2006 Bryan Beaulieu built a 6,000-square-foot (557 sq. m), $2 million solar-and-hydrogen-powered house in Scottsdale, Arizona. The house's solar panels convert sunlight into electricity that powers an electrolyzer. This $50,000 appliance, about the size of a washing machine, separates hydrogen molecules from water. The hydrogen gas is stored in high-pressure tanks and used to power a modified electrical generator built to run on gas. (Beaulieu is planning to add fuel cells in the future.) The generator provides power for appliances, electronics, and lights as well as heat during the winter and air conditioning throughout the long, hot Arizona summers. Hydrogen also fuels the family cars.

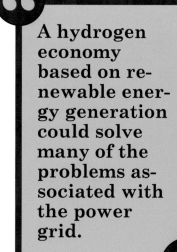

A hydrogen economy based on renewable energy generation could solve many of the problems associated with the power grid.

Beaulieu's house is beautiful as well as energy efficient. It is made up of five hexagonal living pods around a courtyard. The pods are designed to resemble hogans, or traditional Navajo houses. Inside, the ceilings are outfitted with pipes that circulate cool water for air conditioning.

Beaulieu believes his house is a perfect example of what can be achieved through the hydrogen economy. Commenting on his desire to be a model for the future, Beaulieu states: "In a way, this house is a big exhibit. It's Disneyland and NASA."[23]

Free-Range Hydrogen

Beaulieu modeled his home on another southwestern mansion called the Angel's Nest Sustainable Showplace. This multimillion-dollar sustainable energy research and development center was created by Robert Plarr and Victoria Peters just outside Taos, New Mexico. The 8,000-square-foot (743 sq. m) underground home was designed to use biomass, wind, and solar energy to produce what Plarr calls "free-range hydrogen"[24] in electrolyzers. In a 2006 interview, Plarr discussed the roles of wind and solar power in the Angel's Nest hydrogen economy:

> What you have is excess electricity from wind or solar. . . . You have this energy, and you run it in an electrolyzer, and the electrolyzer splits the water, so you use two or three kilowatts to split water to make this hydrogen. . . . What's nice about this hydrogen is that you're storing excess energy. You can store a month's worth if you want depending on how many tanks you put in the ground. If you want three months worth of storage, you could store that excess energy that you're producing onsite.[25]

> "The Angel's Nest hydrogen electrolyzer system uses rainwater collected in storage tanks."

The Angel's Nest hydrogen electrolyzer system uses rainwater collected in storage tanks. The hydrogen is stored in a special cistern, and the distilled water that is produced as exhaust is recycled back into the electrolyzer. Plarr calls this "an elegant, closed energy cycle that cannot harm our world."[26]

Plarr emphasizes that in addition to being environmentally friendly, his home is an insurance policy against disaster and attacks. Because he produces his own hydrogen energy, water, and food, Plarr states, "Our buildings are water resistant, fire resistant, hurricane resistant . . . and if something happens like a power-outage, when you have a sustainable home you're not held hostage by terrorist attacks or natural disasters."[27]

While Plarr's home is more of a dream castle than a typical suburban house, he acts as a consultant to those who are interested in building sustainable homes. Nearly every day, Angel's Nest is visited by architects, educators, engineers, inventors, politicians, scientists, students, and others interested in expanding the hydrogen economy.

The Hydrogen Luxury

Angel's Nest is able to function because Plarr has a great deal of money to spend and a large plot of land to house cisterns, windmills, solar collectors, and other items necessary for hydrogen production. However, for the hydrogen economy to supply the world's energy needs, the system would have to be both practical and inexpensive for use by typical homeowners. And some argue that the science behind hydrogen production and consumption reveals a major problem. Producing hydrogen, even with renewable resources, wastes a great deal of water and electricity.

> " Producing hydrogen, even with renewable resources, wastes a great deal of water and electricity. "

It takes about 20 pounds (9kg) of water, or 2.5 gallons (9.5L), to produce 2.2 pounds (1kg) of hydrogen. The electrolysis to produce the hydrogen consumes 55 kilowatt-hours of electricity. This is enough energy to light a 100-watt lightbulb continuously for 20 days. However, 2.2 pounds of hydrogen when used in a typical fuel cell would only power that lightbulb for little more than 5 days. Therefore, only about 25 percent of the original electrical energy is turned into power, while 75 percent is wasted by electrolysis and the fuel cell.

On a practical level, a homeowner would need four times the number of solar panels or windmills to produce electricity with hydrogen than he or she would need simply to power a home without hydrogen. Using wind turbines as an example, Ulf Bossel, founder of the European Fuel Cell Forum, explains, "If hydrogen is used as the [main source of energy] four wind turbines must be installed to provide these consumers with the same amount of energy. Essentially, only one of these wind turbines produces consumer benefits, while the remaining three are needed to compensate the energy losses arising from the hydrogen luxury."[28]

Most people do not power their homes with solar panels and windmills. And since it is costly, polluting, and wasteful to produce hydrogen with fossil fuels, some consider the hydrogen economy to be unattainable. But Robert Plarr and others like him strongly disagree. Plarr's solar panels generate a surplus of electricity during the day. By turning that extra electricity into hydrogen, Plarr can live without being connected to the power grid. Similar experimental houses have been built in Florida, New Jersey, England, and Malaysia.

> It is costly, polluting, and wasteful to produce hydrogen with fossil fuels.

With the move toward the hydrogen economy, scientists around the globe are searching for more efficient ways to produce and consume the most ubiquitous molecule in the universe. A breakthrough could come at any time that would make the dream of a hydrogen house an everyday reality by the end of the century.

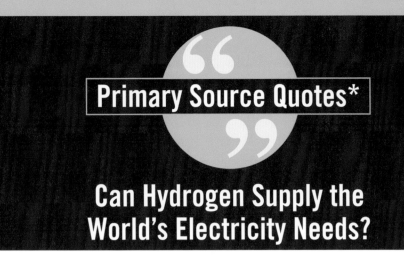

Primary Source Quotes*

Can Hydrogen Supply the World's Electricity Needs?

66A forced transition to a hydrogen economy may prevent the establishment of a sustainable energy economy based on an intelligent use of precious renewable resources.99

—Ulf Bossel, "The Hydrogen Illusion," European Fuel Cell Forum, March/April 2004. www.efcf.com.

Bossel is a fuel cell consultant who heads the European Fuel Cell Forum in Lucerne, Switzerland.

66Wind, solar, water and hydrogen are the elements of sustainable, peaceful existence. They are the enablers of our new coexistence with nature.99

—Robert Plarr, "Angel's Nest—Renewable Energy and Sustainable Architecture," Angel's Nest. www.angels-nest.org.

Plarr is the designer of Angel's Nest, a hydrogen-powered house in Taos.

Bracketed quotes indicate conflicting positions.

* Editor's Note: While the definition of a primary source can be narrowly or broadly defined, for the purposes of Compact Research, a primary source consists of: 1) results of original research presented by an organization or researcher; 2) eyewitness accounts of events, personal experience, or work experience; 3) first-person editorials offering pundits' opinions; 4) government officials presenting political plans and/or policies; 5) representatives of organizations presenting testimony or policy.

66 Four units of energy are thrown away for every one unit of hydrogen energy produced. That is a lot of energy to waste. 99

—Joseph J. Romm, *The Hype About Hydrogen*. Washington, DC: Island, 2005.

Romm is a physicist and executive director of the Center for Energy and Climate Solutions.

66 Hydrogen makes it possible to economically store over time—for the winter season, for example—energy derived from intermittent sources such as solar power. 99

—Peter Hoffman, *Tomorrow's Energy*. Cambridge, MA: MIT Press, 2001.

Hoffman is the editor and publisher of "The Hydrogen & Fuel Cell Newsletter."

66 Free hydrogen does not exist on this planet, so to derive free hydrogen we must break the hydrogen bond in molecules. Basic chemistry tells us that it requires more energy to break a hydrogen bond than to form one. 99

—Adam Grubb, "The Myth of the Hydrogen Economy," *Energy Bulletin*, January 5, 2006. www.energybulletin.net.

Grubb is the founder of the *Energy Bulletin*, an online clearinghouse for information regarding the peak in global energy supply.

"The hydrogen economy has a built-in surcharge. You get back only [25] percent of what you put in. Las Vegas gives you better percentage returns."

—Kenneth S. Deffeyes, *Beyond Oil.* New York: Hill and Wang, 2005.

Deffeyes is a professor emeritus at Princeton University.

....

"Converting wind energy to hydrogen means that it doesn't matter when the wind blows since its energy can be stored on-site in the form of hydrogen."

—Dick Kelly, "Creating Hydrogen," Hydrogen Commerce. http://hydrogencommerce.com.

Kelly is CEO of Xcel Energy, a major supplier of electricity in eight states.

....

"A hydrogen economy . . . means that virtually all military costs and concerns about politically motivated energy supply disruptions would be eliminated."

—Geoffrey B. Holland and James J. Provenzano, *The Hydrogen Age.* Layton, UT: Gibbs Smith, 2007.

Holland is the cofounder of Hydrogen 2000, and Provenzano is the president of Clean Air Now.

....

Facts and Illustrations

Can Hydrogen Supply the World's Electricity Needs?

- It would cost an estimated **$100 billion** to upgrade the transmission lines that carry electricity from power plants to homes and businesses in the United States.

- In 2008 the high-tech company Fujitsu America installed a hydrogen fuel cell to provide **50 percent** of the power to its Sunnyvale, California, campus.

- Fuel cell **gadget chargers** can provide hydrogen power to laptop computers, cell phones, MP3 players, and camcorders.

- An electrolyzer used to separate hydrogen molecules from water costs about **$50,000** and is the size of a washing machine.

- Only **25 percent** of the original electrical energy used to make hydrogen is turned back into electricity.

- In 2008 Korean steelmaker Posco began producing the world's largest fuel cells for **commercial and industrial use**.

- To alleviate a clean water shortage, a power company in Mumbai, India sells **distilled water** that is a by-product of its industrial-sized fuel cells.

- The Westin Hotel in San Francisco uses fuel cells to supply **100 percent** of its electricity requirements, which include providing hot water for rooms and heating for the swimming pool.

- **Platinum** used in most fuel cells is scarce, only produced in **five mines** around the world.

- The Solar-Hydrogen Eco-House in Malaysia, which uses an electro-lyzer to produce hydrogen for the home's appliances, cost less than **$66,000** to build in 2003.

- In the United States **over 400 cell phone towers** use fuel cells as backup power so that phones will continue to work during weather-related emergencies.

Projected Electricity Demand Through 2030

Citizens of the United States will continue to use more electricity due to an expanding population and a growing number of computers, televisions, and other electrical devices. Global warming is expected to add to the problem as hotter weather increases the use of power-hungry air conditioners. This graph from the Energy Information Administration shows the expected increase in electricity demand through 2030.

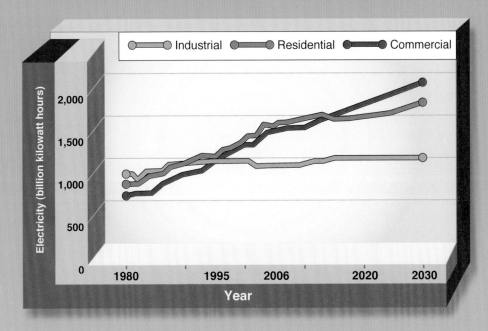

Powering a Home with Renewable Hydrogen

This diagram demonstrates how a home can be powered by hydrogen. Wind turbines or solar panels provide electricity to an electrolyzer, which converts water into hydrogen gas. The gas is stored in a tank underground or in the garage or basement. The hydrogen feeds a fuel cell that produces electricity and water. The electricity provides power for the home while the water can be recycled back into the electrolyzer.

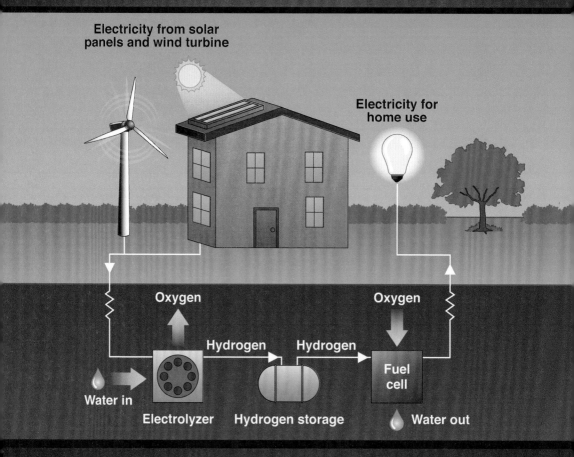

Electricity from solar panels and wind turbine

Electricity for home use

Oxygen

Oxygen

Hydrogen

Hydrogen

Fuel cell

Water in

Electrolyzer

Hydrogen storage

Water out

Source: Rex A. Ewing, *Hydrogen—Hot Stuff, Cool Science: Discover the Future of Energy*, 2007.

How Will Hydrogen Use Impact Global Warming?

> **"Weaning the world away from a fossil-fuel energy regime [with hydrogen] will limit CO_2 emissions . . . and mitigate the effects of global warming on the Earth's already beleaguered biosphere."**
>
> —Jeremy Rifkin, best-selling author and authority on cutting-edge technology.

> **"In the near term, hydrogen is likely to be made from fossil fuel sources, which entails significant greenhouse gas emissions."**
>
> —Joseph J. Romm, physicist and executive director of the Center for Energy and Climate Solutions.

In early 2007 a scientific body called the Intergovernmental Panel on Climate Change (IPCC) issued a report stating that the Earth's climate is rapidly warming. The panel established by the United Nations also stated that there is a 95 percent certainty that the temperature increase is directly traced to human activities that release carbon dioxide, methane, and nitrous oxide into the atmosphere. The IPCC, after studying reports from scientists and climatologists in 177 countries, concluded that there is a 90 percent chance that weather will become more severe. Global warming will cause frequent heat waves, droughts, and excessive rain and snowfall.

The effects of global warming can already be seen. Of the 12 years between 1996 and 2008, 11 were among the warmest on record. Average Arctic temperatures are increasing at nearly twice the average rate, and Arctic ice is thinning at an alarming rate.

Droughts have become more intense, have lasted longer, and have covered larger areas than in the past. And an increase in ocean temperatures has led to more intense hurricanes and cyclones.

In December 2007 the IPCC, along with former vice president Al Gore, won the Nobel Peace Prize, according to the Nobel committee, "for their efforts to build up and disseminate greater knowledge about man-made climate change, and to lay the foundations for the measures that are needed to counteract such change."[29]

> **Average Arctic temperatures are increasing at nearly twice the average rate.**

Hydrogen Power's Potential

The 2007 IPCC report was among dozens of studies that have documented the causes and effects of global warming during the past decade. The studies have largely concluded that there is an overabundance of carbon dioxide in the atmosphere that is trapping too much of the sun's heat on Earth. This has caused average global temperatures to rise about 1.5°F (0.84°C) in the past century. Experts predict the mean temperature will continue to rise 1.2 to 6 degrees in the twenty-first century, creating serious environmental problems for the planet.

Most climatologists agree that the only way to slow global warming is for humanity drastically to reduce carbon dioxide emissions from 60 to 80 percent of current levels. Because most CO_2 is produced by burning fossil fuels, some believe the best way to lower emissions is to move toward a hydrogen economy within the next 50 years; however, the hydrogen would have to be produced using methods that do not add CO_2 to the atmosphere.

Iceland's Hydrogen Experiment

The dramatic effects of global warming and the possible benefits of hydrogen power may already be seen on the small island nation of Iceland. Located just south of the Arctic Circle in the North Atlantic Ocean, the stark beauty of Iceland includes icy glaciers, snow-capped mountain ranges, and rushing, snow-fed rivers. But in the past 20 years, average temperatures around the Arctic have increased by about 1.5°F (0.84°C). This has accelerated the melting of Iceland's ice caps, glaciers, sea ice, and frozen soil (called permafrost).

In addition to its frozen beauty, Iceland also contains numerous geysers. Icelanders have harnessed the intense heat from these geysers to produce electricity in geothermal power plants. Combined with hydroelectric dams on the island's cascading rivers, Iceland's residents enjoy cheap electric power that produces no carbon dioxide emissions. Yet 30 percent of total energy consumption in Iceland still comes from oil, which is primarily used to power cars, trucks, and boats.

In order to reduce global warming and free the nation of its reliance on oil, Iceland has taken steps to create a carbon dioxide–free hydrogen economy. The center of this experiment is the Shell Oil hydrogen refueling station opened in 2003 in the capital city of Reykjavik. The station produces hydrogen from water by electrolysis using electricity generated from geothermal sources. From 2003 to 2008 the government used the station to power 3 experimental Daimler-Benz buses equipped with hydrogen fuel cells. In 2008 Hertz Car Rental began offering customers Toyota Priuses that were modified to run on hydrogen. The cars, which charge their batteries with internal combustion engines, burn hydrogen instead of gas. They can be rented for $300 a day.

The Shell hydrogen station also produces fuel for a hydrogen ship, the first such craft in the world. The *Elding*—Icelandic for "lightning," uses a fuel cell to run an engine that powers its lights. For about $70 tourists can board the ship for whale-watching tours. When whales come into view, the crew shuts down the ship's main power plant. Without the noise of the diesel engine, people can hear the noises the giant mammals make as they swim through the sea and shoot jets of water from their blowholes.

> Some believe the best way to lower emissions is to move toward a hydrogen economy within the next 50 years.

In a venture between private industry, researchers, and the government, the *Elding*'s diesel engine will eventually be converted to run on hydrogen. Commenting on these plans, ship owner Vignir Sigursveinsson says, "When we have the hydrogen machine, the boat will be completely soundless, which will make the experience of seeing the whales in their natural habitat even more magical."[30]

Within the next decade, Iceland wants to convert its entire fishing fleet, one of the world's largest, to run on hydrogen. But while the CO_2-free hydrogen economy looks promising in Iceland, problems remain. A fishing ship would need a huge hydrogen storage tank to make long trips at sea. And hydrogen-powered cars, trucks, and buses are impossible to find for public sale, even if people wanted to pay the high prices manufacturers would charge for such vehicles. However, if these issues can be resolved, the 300,000 residents of Iceland could lower their carbon dioxide emissions to near zero through hydrogen power. And government experts believe that this may happen within the next few decades.

Hydrogen from Coal

Without major sources of geothermal energy, it would be nearly impossible for the United States to reproduce Iceland's experiment. However, industrialized nations have found other ways to produce limited amounts of CO_2-free hydrogen without adding to global warming. One of these techniques involves coal.

> "
> **Iceland's residents enjoy cheap electric power that produces no carbon dioxide emissions.**
> "

In 2008 about 18 percent of the world's hydrogen was derived from coal. It was produced in coal power plants by what is called an Integrated Gasification Combined Cycle (IGCC) system. This technology, also called coal gasification, transforms coal into a synthetic gas, or syngas, made up of hydrocarbons. The syngas is then treated with steam, which "cracks," or separates, the gas into molecules of hydrogen and carbon dioxide. The stream of pure hydrogen is used to fire electrical turbines in the plant or sold commercially.

Capturing and Storing Carbon Dioxide

The carbon dioxide produced during IGCC is released into the atmosphere, adding to global warming. However, researchers are working on a method called carbon capture and storage (CCS), or carbon sequestration. These terms describe ways to burn coal and create hydrogen without adding to global warming. In the CCS process, the CO_2 produced by the power plant is injected under high pressure into large underground caverns. These geological formations may include depleted oil and natural gas fields, coal seams, or salt caverns. There are also methods to sequester the carbon dioxide by pumping it deep into the ocean, creating "lakes" of CO_2 beneath the waves.

Carbon capture and storage reduces CO_2 emissions from a typical coal power plant by 80 to 90 percent. However, capturing, pressurizing, and transporting carbon dioxide through pipelines requires a great deal of energy. Under normal circumstances it takes four units of energy to produce one unit of hydrogen. If a coal gasification plant used CCS methods, it would take six to eight units to produce a single unit of hydrogen.

Experiments with CCS and hydrogen-producing coal power plants are taking place in Algeria, Canada, China, Germany, the Netherlands, and the United States. But environmentalists oppose IGCC plants for several reasons. Even with CCS technology, coal-burning power plants continue to emit dangerous pollutants such as mercury and sulfur oxide, which is responsible for acid rain. In addition, great quantities of water are used in the IGCC process. This not only wastes precious freshwater supplies, but the water is fouled by chemicals in the process, adding to pollution problems. And many environmental activists believe that there should be a moratorium on building coal plants because CCS technology has not been perfected. As Emily Rochon, climate and energy campaigner at Greenpeace International, states: "Carbon capture and storage is . . . the ultimate coal industry pipe dream. Governments and businesses need to reduce their emissions—not search for excuses to keep burning coal."[31]

Despite resistance by environmentalists, coal remains cheap and plentiful—the United States has enough to take care of its energy needs at current rates for more than 250 years. Therefore, researchers continue searching for "clean" methods to burn coal while converting the hydrocarbons into hydrogen.

> " The 300,000 residents of Iceland could lower their carbon dioxide emissions to near zero through hydrogen power. "

Nuclear Hydrogen

Production of hydrogen through electrolysis is much less complicated than coal gasification technology because electrolysis only requires electricity.

Therefore, proponents of the hydrogen economy such as former president George W. Bush believe that nuclear power should play an important role in hydrogen production. In 2008 there were 442 nuclear power plants operating throughout the world. The United States had 104 of those reactors, and they produced about 20 percent of the nation's electricity while emitting zero carbon dioxide.

The enriched uranium that fuels nuclear power plants is incredibly powerful. The energy from a single pound of nuclear fuel contains the equivalent of 250,000 gallons (946,353L) of gasoline, according to the U.S. Department of Energy. But while this energy produces no CO_2, nuclear power is not as clean as proponents suggest. A large nuclear reactor generates several tons of radioactive waste each year. This waste is highly toxic to humans and the environment and remains so for over 30,000 years. The United States currently has over 60,000 tons (54,431 metric t) of uranium waste. Most of it is stored at 68 sites around the country, mainly at nuclear power plants where it is generated. While the United States has spent billions to create the Yucca Mountain Repository in Nevada to store the waste, the site has been plagued with problems. It is not expected to become operational until 2030, and even then, critics say the site will leak radioactive waste.

> " Capturing, pressurizing, and transporting carbon dioxide through pipelines requires a great deal of energy. "

Because of the problems associated with nuclear waste disposal, environmentalists oppose building more nuclear reactors to produce a CO_2-free fuel. As renewable energy expert Rex A. Ewing writes, the nuclear industry "generates several tons of highly-lethal radioactive waste every year, and there is no general agreement on what to do with it. Until that issue is resolved, nuclear energy is likely to remain mired in a bog of skepticism."[32]

If nuclear power was used to create enough hydrogen to fuel American transportation needs, the amount of radioactive waste would double. And it would take more than 100 new nuclear plants to make that amount of hydrogen. Since it can take 10 to 20 years to bring a new nuclear plant online, this method of generating CO_2-free hydrogen is not feasible in the short run. Scientists are developing a new type of nuclear power plant that creates hydrogen as a by-product, but this technology will not be ready for at least a decade.

A Growing Role

Every method of hydrogen production has its drawbacks. For example, it would take solar panels covering an area the size of North Dakota to produce enough hydrogen to eliminate reliance on fossil fuels for transportation. But proponents believe that clean-burning hydrogen fuel is the best way to slow global warming. Whether it is generated by clean coal, nuclear power, or renewable energy, hydrogen will likely play a growing role in the efforts to keep the planet from overheating.

> " A large nuclear reactor generates several tons of radioactive waste each year. "

Primary Source Quotes*

How Will Hydrogen Use
Impact Global Warming?

❝If hydrogen is to play a significant role in the future energy markets, it must be made from carbon-free or carbon-neutral energy sources.❞

—Venki Raman, "Linking Hydrogen and Renewable Energy Strategies Is Vital," National Hydrogen Association. www.hydrogenassociation.org.

Raman is president of Protium Energy Technologies, which promotes hydrogen and fuel cell technology.

❝Producing hydrogen from coal . . . deserves serious consideration because the costs of coal are lower and are likely to remain so even with significant increases in demand.❞

—Daniel Sperling and James S. Cannon, *The Hydrogen Energy Transition.* Burlington, MA: Elsevier, 2004.

Sperling is an engineer and an expert on hydrogen transportation, and Cannon is the president of Energy Futures, Inc.

Bracketed quotes indicate conflicting positions.

* Editor's Note: While the definition of a primary source can be narrowly or broadly defined, for the purposes of Compact Research, a primary source consists of: 1) results of original research presented by an organization or researcher; 2) eyewitness accounts of events, personal experience, or work experience; 3) first-person editorials offering pundits' opinions; 4) government officials presenting political plans and/or policies; 5) representatives of organizations presenting testimony or policy.

❝I challenge our nation to commit to producing 100 percent of our electricity from renewable energy and truly clean carbon free sources within 10 years.❞

—Al Gore, "Can We Save the Planet and Rescue the Economy at the Same Time?" *Mother Jones*, November/December 2008.

Gore is an environmental activist and a former vice president of the United States.

....

❝While some day we may be able to produce hydrogen by breaking up water molecules ... through renewable energy technologies, right now the most cost-effective way to produce hydrogen is with coal.❞

—Chris Shaddix, "Coal for Hydrogen: Experiments Examine Hydrogen-Production Benefits of Clean Coal Burning," Physorg.com, April 04, 2006. www.physorg.com.

Shaddix is the principal researcher for clean coal combustion at Sandia Corporation's Combustion Research Facility.

....

❝Clearly the day will come when we have to wean ourselves away from our coal habit, and most would say the sooner the better.❞

—Rex A. Ewing, *Hydrogen—Hot Stuff, Cool Science: Discover the Future of Energy*. Masonville, CO. PixyJack, 2007.

Ewing is a writer who specializes in renewable energy topics.

....

66 The most promising option for avoiding CO_2 is to transform fossil fuels to hydrogen and then use this [as] fuel. 99

—Bent Sørensen, *Hydrogen and Fuel Cells*. Burlington, MA: Elsevier, 2005.

Sørensen is a professor of physics at Roskilde University in Denmark.

66 As it is essential to reduce the global use of fossil fuels, it is important to explore the feasibility of . . . nuclear energy to produce hydrogen. 99

—Masao Hori, "Hydrogen from Nuclear Power," *21st Century*, Spring 2006.

Hori is a nuclear physicist.

66 I am skeptical that nuclear-generated hydrogen could be a practical solution . . . [to global warming] since 100 or more nuclear plants would be needed to replace a significant fraction of U.S. transportation fuel with hydrogen. 99

—Joseph J. Romm, *The Hype About Hydrogen*. Washington, DC: Island, 2005.

Romm is a physicist and executive director of the Center for Energy and Climate Solutions.

How Will Hydrogen Use Impact Global Warming?

- The Intergovernmental Panel on Climate Change (IPCC) says there is a **95 percent** certainty that global warming is caused by human activities.

- Carbon dioxide emissions from human activities account for **70 percent** of global warming pollution. Methane from landfills and cattle flatulence makes up another **24 percent**.

- Atmospheric CO_2 concentrations from burning fossil fuels have increased **75 percent** since the 1980s.

- By using a 700W fuel cell for electricity, a family of 4 can reduce its CO_2 emissions by **30 percent** or 1 ton per year.

- **Eleven percent** of Iceland is covered in glaciers, but these giant ice sheets are expected to disappear by 2100 due to global warming.

- A power plant using carbon capture and storage technology burns **25 to 40 percent** more coal than a non-CCS plant.

- In order to reduce the CO_2 emissions associated with offshore oil production in Norway, fuel cells are being used to provide **electricity on oil drilling platforms**.

- In an effort to slow global warming, the European Union established a **Hydrogen and Fuel Cell Technology commission** in 2005 to develop and deploy fuel cells in home, factories, and government buildings.

The Greenhouse Effect: Natural and Human-Made

The greenhouse effect is a natural part of the Earth's warming cycle caused by the sun's rays hitting the planet. As this graph shows, (1) the sunlight strikes the Earth and its energy is converted into infrared radiation, or heat. About 70 percent of the sun's energy is absorbed by the Earth's surface, and the rest bounces back into the atmosphere, (2) which is made up of water vapor and tiny amounts of carbon dioxide, methane gas, and other components. The atmosphere traps some of the infrared radiation generated by the sun (3) and sends it back to Earth (4). The rest of the heat escapes into space (5). However, greenhouse gases created by human activity (6) increase the amount of heat absorbed by the atmosphere and cause an increase in average global temperatures.

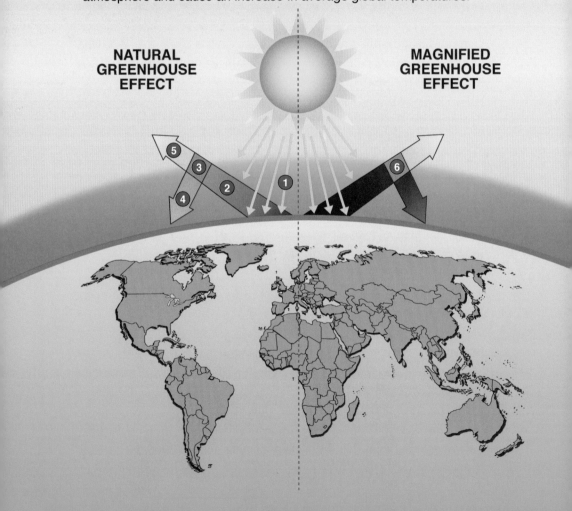

NATURAL GREENHOUSE EFFECT

MAGNIFIED GREENHOUSE EFFECT

Source: The Pew Center on Global Climate Change, "The Causes of Global Climate Change," August 2008.

- In 2008 Barack Obama spoke of creating **5 million "green collar" jobs** in industries that produce solar panels, wind turbines, hydrogen electrolyzers, fuel cells, and fuel cell vehicles.

- The most cost-effective way to produce hydrogen is with **coal**.

- Nuclear power plants can **convert seawater** into CO_2-free hydrogen fuel and potable water.

Sources of Greenhouse Gas

Carbon dioxide, a by-product of burning fossil fuels, is responsible for 70 percent of all greenhouse gases. Methane from landfills and cattle flatulence constitutes another 24 percent. Nitrous oxide generated by the agricultural and chemical industries accounts for the remaining 6 percent of global warming gases.

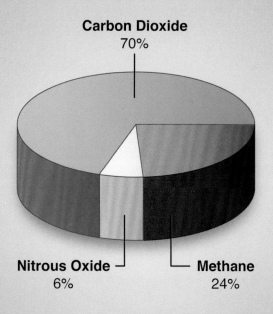

Carbon Dioxide
70%

Nitrous Oxide
6%

Methane
24%

Source: Paul Brown, *Global Warning.* 2007.

- After spending **$50 million**, the U.S. Department of Energy canceled the proposed **$1.8 billion** FutureGen clean coal and hydrogen production plant, calling it too expensive.

Carbon Dioxide Levels Continue to Rise

Researchers began measuring atmospheric carbon dioxide concentration in 1960 at the Mauna Loa Observatory in Hawaii. Since that time, the amount of carbon dioxide, measured in parts per million, has climbed steadily. At about 385 parts per million in 2008, carbon dioxide levels are higher than at any time in the previous 420,000 years.

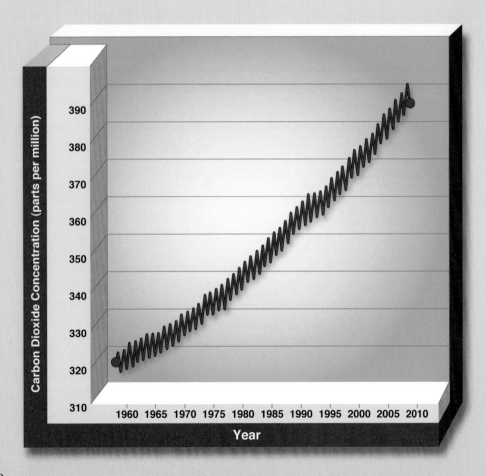

Source: Scripps CO_2 Program, "Mauna Loa Record," 2008. http://scrippsco2.ucsd.edu.

Hydrogen Produces No Carbon Dioxide

When a car burns eight gallons of gasoline it produces about 155 pounds of carbon dioxide. This graph shows how much CO_2 is emitted by other fuel sources when producing the energy equivalent of eight gallons of gas.

To produce 1 million BTUs, or the energy contained in 8 gallons of gasoline:

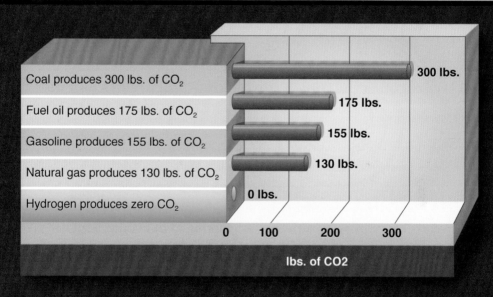

Coal produces 300 lbs. of CO_2 — 300 lbs.

Fuel oil produces 175 lbs. of CO_2 — 175 lbs.

Gasoline produces 155 lbs. of CO_2 — 155 lbs.

Natural gas produces 130 lbs. of CO_2 — 130 lbs.

Hydrogen produces zero CO_2 — 0 lbs.

lbs. of CO2

Source: Joseph David Cohen, "Manufacturing Hydrogen for Our Clean Energy Future," HydrogenHighway.com, 2006. www.hydrogenhighway.com.

What Is the Future of Hydrogen Power?

66 Hydrogen is the future of energy and the future is now.99

—Robert Orr Jr. is the author of *Hydrogen Future Now.*

66 Hydrogen is the fuel of the future—and always will be.99

—Frank Marcus, automotive engineer, test driver, and technical director for *Car and Driver* magazine.

In 1766 British scientist Henry Cavendish was the first to recognize the powerful energy generated by pure hydrogen, labeling it "inflammable air." In the twentieth century hydrogen was used to power a small number of automobiles, airplanes, and spacecraft. But in the twenty-first century, scientists are attempting to fit hydrogen into an energy scheme Cavendish could never have imagined. They want hydrogen to replace completely the fossil fuels that have been driving industrial society since the late eighteenth century. And in this role, "inflammable air" has yet to meet the challenges.

Although fuel cells today power cars, trucks, buses, motorcycles, homes, and offices, even hydrogen's strongest proponents speak of the fuel in the future tense. Fuel cell automobiles might make up 2 percent of all cars by 2015. Hydrogen fueling stations could be everywhere by 2025. And, envisioning the future, Holland and Provenzano write, "The Hydrogen Age is fully manifested [in 2050]. The world runs almost entirely on hydrogen and electricity. [Human-made] CO_2 production has dropped to a level not seen since the beginnings of the industrial revolution. Atmospheric CO_2 levels are falling slowly but steadily."[33]

Of course such projections are entirely speculative. For the hydrogen age to fall into place, massive sums of money will have to be invested by government and industry. For example, the National Research Council, the federal advisory committee on science, medicine, and engineering, reported that it will take $55 billion in government funds in the next 15 years for hydrogen vehicles to be competitive with traditional cars by 2050.

Consumers must also be willing to sacrifice if the hydrogen economy is to take hold. The move will undoubtedly raise prices for energy and transportation, if only for the short term. And those who work for automakers, coal-mining operations, trucking firms, railroads, petroleum producers, and oil refiners will feel major impacts on their livelihoods. Thousands of jobs will simply disappear, and others will move to different parts of the country. In fact, the entire face of civilization, which was built on coal, oil, natural gas, and gasoline over a period of several centuries, would have to be transformed in a few short decades.

Cheaper Electrolyzers

Hydrogen proponents understand that the move to the hydrogen age must have broad public support and that the transition should create as few economic disruptions as possible. Therefore, scientists are working on several exciting projects that focus on ways to reduce the high cost of hydrogen production.

In 2008 researchers at the Massachusetts Institute of Technology (MIT) discovered a method that could drastically lower the price of electrolysis. Traditional electrolyzers use expensive platinum as a catalyst to separate oxygen and hydrogen molecules in water. But experiments conducted by Daniel Nocera, the Henry Dreyfus Professor of Energy at MIT, have shown that cobalt and phosphate may be used as a catalyst. This cobalt-phosphate mixture separates the molecules in water using nothing but abundant, nontoxic

> " Even hydrogen's strongest proponents speak of the fuel in the future tense. "

natural materials. The price is markedly lower than platinum, since cobalt costs about $2.25 an ounce (28g) and phosphate costs about 5 cents an ounce. And unlike complicated conventional electrolyzers, the new method is simple. As Nocera says, "It's easy to set up. We figured out a way just using a glass of water at room temperature."[34]

Nocera believes that by 2020 his discovery will make electrolyzers much more affordable for homeowners. In addition, by not using platinum, extremely large electrolyzers could be built to produce fuel for millions of hydrogen-powered vehicles. Like the hydrogen economy envisioned by Bockris in the 1960s, these huge hydrogen generators could be centrally located in sunny rural areas, such as deserts. Far from major cities, massive fields of solar collectors would be employed to provide electricity for hydrogen production.

> " The entire face of civilization, which was built on coal, oil, natural gas, and gasoline over a period of several centuries, would have to be transformed in a few short decades. "

Pebble-Bed Nuclear Generators

A much more complicated method for producing hydrogen concerns a new generation of nuclear reactors currently under development in China and South Africa. These generators are called pebble-bed reactors (PBRs) or pebble-bed modular reactors (PBMRs). They are so named because instead of using conventional fuel rods of enriched uranium, PBRs use tennis ball–sized pebbles with uranium cores.

PBRs are relatively small. They are about one-fifth the size of traditional nuclear reactors and can be sited on an area about half the size of a football field. These plants are also inexpensive because they can be assembled quickly from parts that will be mass-produced in a factory.

Pebble-bed reactors are designed to eliminate the possibility of meltdowns, one of the greatest dangers associated with nuclear power plants. For example, in 1979 one of the nuclear reactors at Three Mile Island in

Harrisburg, Pennsylvania, experienced a partial meltdown. Hundreds of thousands of people were forced to evacuate the area, and fear of nuclear disaster swept the country while scientists took several days to solve the problem. In 1986 a major explosion at the Chernobyl power plant in Ukraine released a toxic plume of radioactive fallout that drifted across Europe. In the aftermath, about 9,000 people became severely ill or died as a result.

To prevent such disasters, pebble-bed reactors are cooled with helium instead of water, as in traditional reactors. Helium is noncorrosive and will not react with anything else in nature. Therefore, it cannot become radioactive. If helium leaked from a PBR, it would not poison the surrounding community. In addition, PBRs are designed to be passively safe, which means that even without the help of high-pressure helium gas used for cooling, the nuclear reactions within the PBRs stabilize and will not overheat. The plants do produce radioactive waste, however.

> " By not using platinum, extremely large electrolyzers could be built to produce fuel for millions of hydrogen-powered vehicles. "

In addition to producing electricity, PBRs can be equipped to crack water into hydrogen and oxygen. Hydrogen can then be piped into local communities for use in fuel cells or modified internal combustion engines. Supporters of pebble-bed reactors also point out that the plants will help fight global warming. The energy content in one tennis ball–sized pebble of enriched uranium is equal to about 3,500 pounds (1,588kg) of coal, which would produce 10,000 pounds (4,536kg) of carbon dioxide if burned. Considering that each PBR contains about 440,000 pebbles, this represents a significant reduction in greenhouse gases.

The Nuclear Hydrogen Initiative

In the United States, former president George W. Bush proposed the Nuclear Hydrogen Initiative (NHI) in 2003. The goal of the NHI is to fund research aimed at producing commercial hydrogen in newly de-signed nuclear power plants by 2019. The plants are called Generation IV nuclear reactors because they evolved from the Generation III plants built in the 1970s and 1980s. According to the Department of Energy Web site, "If successful, this research could lead to a large-scale, emission-free, domestic hydro-gen production capability to fuel a future hydrogen economy."[35]

> **Pebble-bed reactors are designed to eliminate the possibility of melt-downs, one of the greatest dangers associated with nuclear power plants.**

Older nuclear plants produce a great deal of heat that goes unused. The new gen-eration of nuclear plants, expected to come online in 2020, would use this excess heat to crack water into hydrogen and oxygen. In a process called thermochemical water splitting, water is heated to about 1,382°F (750°C). At this temperature, about twice as much hydrogen can be produced as through traditional electrolysis.

Generation IV nuclear plants will pro-duce CO_2-free hydrogen at double the cur-rent efficiency. But environmentalists point out that the Bush initiative plans to derive 90 percent of its hydrogen from nonrenewable resourc-es. To fight this aspect of the initiative, eight of the nation's leading environmental, consumer, and public policy organizations, including Friends of the Earth, Greenpeace, the League of Conservation Voters, Public Citizen, and the Sierra Club, formed the Green Hydrogen Co-alition. This group criticized the Bush administration for promoting "black" hydrogen made with fossil fuels and nuclear power, rather than "green" hydrogen made with renewable energy sources such as solar cells, wind turbines, and biomass.

Hydrogen from Pond Scum

The greenest hydrogen of all comes from pond scum, which cracks water molecules naturally. Technically, the bright green algae is called *Chlamydomonas reinhardtii*, and researchers discovered more than 60 years ago that the aquatic microorganisms could make trace amounts of hydrogen under certain natural conditions. In 2001 University of California–Berkeley professor Tasios Melis perfected a laboratory method that forces is the algae to generate large quantities of hydrogen on a continuous basis.

Melis's discovered that the algae must consume minute amounts of sulfur to survive. When the microorganisms are deprived of sulfur, they starve. This causes their internal production of protein to stop, and they begin producing hydrogen, which allows them to live for a time without negative effects.

The hydrogen production is actually solar powered because algae collect energy from the sun through photosynthesis like every other green plant. As the microorganisms grow, the energy is stored as carbohydrates and other fuels. In a process called "biophotolysis" or "photobiological hydrogen production," Melis transfers the algae, which looks like lime green soda, into 1-quart (about 1L) glass bottles where no sulfur is present. After 24 hours in the closed bottles, the cells consume all the oxygen and begin producing hydrogen, which bubbles to the top and is bled off into tall glass tubes. Commenting on the process, Melis says: "It was actually a surprise when we detected significant amounts of hydrogen coming out of the culture. We thought we would get trace amounts, but we got bulk amounts."[36]

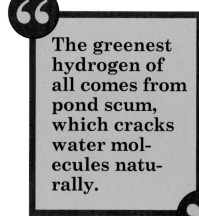

The greenest hydrogen of all comes from pond scum, which cracks water molecules naturally.

After four days of generating hydrogen, the algae must be poured out of the bottles and returned to sunlight, where they can regenerate through photosynthesis. After several days, it can be tapped once again to generate hydrogen.

In experiments, Melis has been able to produce up to 1 quart (1L) of hydrogen per hour using a large tank called a bioreactor filled with 132 gallons (500L) of water and algae. Although he has yet to produce commercial quantities of hydrogen, Melis is working with various government agencies to continue his research. He is convinced that a small pond of algae could produce enough hydrogen to run a dozen cars and that the algae process can be conducted using seawater. However, an area about the size of New Mexico would be needed to grow enough algae to satisfy all the current energy demands for the United States. But as Melis points out, coal mines, oil and natural gas fields, and power plants all consume a great deal of real estate. He figures that 12 percent of the Sonoran Desert between California and Arizona covered in seawater and algae could replace a large percentage of the petroleum used in the United States. This type of hydrogen production would be completely friendly to the environment, requiring only water, sunlight, and green algae. As such, hydrogen from algae is one of the cleanest sources of power ever devised.

The Hydrogen Energy Web?

If production technology can be perfected, experts say, hydrogen might be able to replace fossil fuels by 2100. This would fundamentally change the market for electricity. Consumers with individual fuel cells will be able to produce their own electricity and share it with others by connecting to a nationwide electrical grid called the hydrogen energy web. This would transform consumers from being just passive energy users to also being energy producers. This would completely change the roles of large power companies, turning them into "virtual power plants," according to Jeremy Rifkin: "Their new mandate would be to manufacture and market fuel cells . . . and coordinate the flow of energy over the existing power grids."[37]

> An area about the size of New Mexico would be needed to grow enough algae to satisfy all the current energy demands for the United States.

Despite the optimistic predictions by Rifkin and other proponents, the hydrogen age has been predicted for centuries. When science-fiction writer Jules Verne published a book called *Mysterious Island* in 1874, one of his characters described a fantastic future: "I believe that water will one day be employed as fuel, that hydrogen and oxygen which constitute it . . . will furnish an inexhaustible source of heat and light, of an intensity of which coal is not capable. . . . When the deposits of coal are exhausted we shall heat and warm ourselves with water. Water will be the coal of the future."[38]

Just when Verne's prophecy will actually become reality is impossible to predict. Some have been preparing for a hydrogen future for decades, but the dream of producing pure power from water remains elusive and unpredictable.

What Is the Future of Hydrogen Power?

"Folks who claim to know how our system works . . . have told us we might as well forget about our political system doing anything bold, especially if it is contrary to the wishes of the special interests."

—Al Gore, "Can We Save the Planet and Rescue the Economy at the Same Time?" *Mother Jones*, November/December 2008.

Gore is an environmental activist and a former vice president of the United States.

....

"I'll invest $150 billion over the next decade in affordable, renewable sources of energy."

—Barack Obama, "Barack Obama's Acceptance Speech," *New York Times*, August 28, 2008. www.nytimes.com.

Barack Obama is the forty-fourth president of the United States.

....

Bracketed quotes indicate conflicting positions.

* Editor's Note: While the definition of a primary source can be narrowly or broadly defined, for the purposes of Compact Research, a primary source consists of: 1) results of original research presented by an organization or researcher; 2) eyewitness accounts of events, personal experience, or work experience; 3) first-person editorials offering pundits' opinions; 4) government officials presenting political plans and/or policies; 5) representatives of organizations presenting testimony or policy.

❝The energy challenges our country faces are severe and have gone unaddressed for far too long. Our addiction to foreign oil doesn't just undermine our national security and wreak havoc on our environment—it cripples our economy and strains the budgets of working families all across America.❞

-- TheWhite House, "The Agenda: Energy and the Environment," January 2009. www.whitehouse.gov.

The White House Web site presents the Obama administration's positions on energy and many other issues.

❝[The] shift to hydrogen could be compared with the shift from coal to oil in the 19th century. . . . [The] shift to fuel cells could be as significant as the introduction of electricity 100 years ago.❞

—Chris Borroni-Bird, "Hydrogen Power: A Discussion with Chris Borroni-Bird," Scientific American Frontiers, PBS. www.pbs.org.

Borroni-Bird designs concept cars for General Motors and is a leading expert on fuel cell vehicles.

❝The future will certainly not be a simple replacement of fossil fuels by hydrogen, but a more complex substitution process involving physical and chemical energy carriers.❞

—Ulf Bossel, "We Need a Renewable Energy Economy, Not a Hydrogen Economy," European Fuel Cell Forum, September 2003. www.efcf.com.

Bossel is a fuel cell consultant who heads the European Fuel Cell Forum in Lucerne, Switzerland.

"Construction of the infrastructure needed for our hydrogen based future will produce lifelong jobs for a huge portion of the world's population."

—Joseph David Cohen, "Manufacturing Hydrogen for Our Clean Energy Future," HydrogenHighway.com, May 14, 2004. www.hydrogenhighway.com.

Cohen is a retired engineer who worked for GE Aircraft Engines.

"If we develop hydrogen power to its full potential, we can reduce our demand for oil by over 11 million barrels per day by the year 2040."

—George W. Bush, "Hydrogen Fuel Initiative Can Make 'Fundamental Difference,'" The White House, February 6, 2003. www.whitehouse.gov.

Bush was the forty-third president of the United States.

"[All] the cars in Britain could be converted to run on fuel cells using hydrogen from electrolyzers and using electricity obtained only from wind turbines, . . . in 20 years."

—Rob Thring, "Hydrogen, Not Biofuels, Will Power the Future," *Times Online*, April 28, 2008. www.timesonline.co.uk.

Thring is a professor of fuel cell engineering at Loughborough University in the United Kingdom.

❝Carmakers and researchers hoping to save the planet . . . have to parade these fuel cells and hydrogen cars in front of us so we all feel that someone is doing something about the problem.❞

—Mark Vaughn, "Hydrogen Is Bunk," *AutoWeek,* August 13, 2007. www.autoweek.com.

Vaughn is the senior editor for *AutoWeek*.

❝To get serious about energy policy, America needs to abandon, once and for all, the false promise of the hydrogen age.❞

—Robert Zubrin, "The Hydrogen Hoax," *New Atlantis*, Winter 2007. www.thenewatlantis.com.

Zubrin is an aerospace engineer and the president of Pioneer Astronautics, a research and development firm.

❝Imagine a world running on hydrogen later in this century: Environmental pollution will no longer be a concern. . . . Personal transportation will be cheaper to operate and easier to maintain. . . . Life will get better.❞

—Spencer Abraham, "Remarks by Energy Secretary Spencer Abraham at the Global Forum on Personal Transportation," U.S. Department of Energy, November 12, 2002. www.eroei.com.

Abraham was U.S. secretary of energy from 2001 to 2005.

What Is the Future of Hydrogen Power?

- It will take **$55 billion** in government funds during the next 15 years for hydrogen vehicles to be competitive with traditional cars by 2050.

- In 2006 South Korea established a three-stage goal to switch the nation to a **completely hydrogen-powered economy by 2040**.

- **Generation IV nuclear plants** will produce CO_2-free hydrogen at double the efficiency of traditional electrolyzers.

- China plans to build more than **200** Generation IV nuclear reactors by 2050.

- In 2008 aerospace company Boeing helped found the Algal Biomass Organization to develop methods for using **algae to produce hydrogen jet fuel**.

- Fuel cell vehicles of the future will produce **20 kilowatts of power**. When parked they will be able to supply the electricity needs of **12 typical homes**.

- A small **pond of algae** can produce enough hydrogen to run a dozen cars, and the process can be conducted using seawater.

- If every car on earth was powered by hydrogen there would be a huge increase in the amount of **water vapor in the air** which might increase the frequency of **severe rainstorms** in populated areas.

- A pebble bed nuclear reactor uses **440,000** tennis-ball sized graphite pebbles with uranium cores to produce electricity and hydrogen.

- Chevron is investing in **microwave technology** that may be used someday to break down old tires, plastic bottles, and other trash into hydrogen gas and carbon powder.

Americans Concerned About Energy and Pollution

In 2007, Gallup asked Americans what they think will be the most important issues facing the United States in 25 years. As this graph shows, 22 percent expressed concern that the environment, pollution, and lack of energy sources will be major problems.

Looking ahead, what do you think will be the most important problem facing our nation 25 years from now?

Problem	2007	2006	2005	2004	2003	2002	2001
Environment/pollution	14%	8%	6%	8%	9%	10%	11%
Lack of energy sources/energy crisis	8%	10%	5%	2%	8%	5%	7%
Social Security	8%	10%	23%	10%	5%	6%	8%
Healthcare	8%	7%	6%	6%	6%	5%	3%
Terrorism	7%	3%	4%	4%	5%	5%	–%
Economy in general	5%	8%	9%	12%	14%	12%	5%

Source: Gallup Poll, "Environmental Concerns Hold Firm During Past Year," March 26, 2007. www.gallup.com.

Moving Toward a Hydrogen Economy

This graph shows how the United States government plans to move toward a hydrogen economy in the future. Until 2015, research and development will be paid for by government programs. Between 2010 and 2025, industry will coordinate with government to commercialize a new generation of fuel cells while investing in hydrogen infrastructure. Between 2015 and 2035, markets will expand as hydrogen power and transportation systems are made widely available. Finally, between 2025 and 2045, the hydrogen economy will be expanded to all regions.

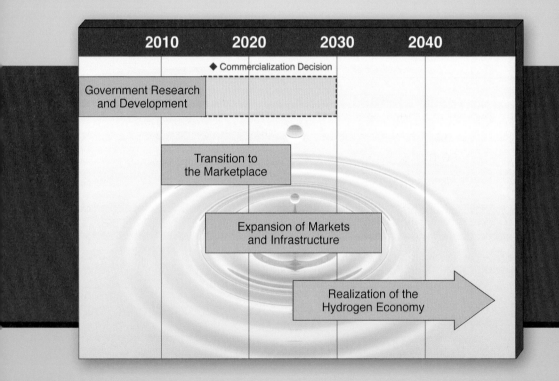

Source: Hydrogen.gov, "Realizing the Hydrogen Economy," September 23, 2006. www.hydrogen.gov.

Half of Americans Do Not Want More Nuclear Power Plants Built

In 2003, President George W. Bush instituted the Nuclear Hydrogen Initiative to fund research aimed at producing commercial hydrogen in newly designed nuclear power plants by 2019. However, this 2008 Harris poll found that nearly 50 percent of all Americans are opposed to construction of any new nuclear power plants in the United States.

Q: How much do you favor or oppose building new nuclear power plants?

FAVOR	**52%**
Strongly favor	20%
Favor more than oppose	32%
OPPOSE	**48%**
Oppose more than favor	31%
Strongly oppose	17%

Note: Percentages do not add up to 100 percent due to rounding.

Source: Harris Interactive, "Adults in Five Largest European Countries and the U.S. Supportive of Renewable Energy, but Unwilling to Pay Much More for It," February 26, 2008.

- In 2009, General Motors and Honda were developing hydrogen fueling stations that could be installed at **large retailers, shopping malls, new car dealerships, and supermarkets.**

Nations Developing Generation IV Nuclear Energy Systems

The Generation IV International Forum (GIF) was established in 2000 for member nations to develop new nuclear power systems for meeting future energy challenges. As this map shows, countries in the GIF include the United States, Brazil, Canada, France, Japan, South Korea, South Africa, and the United Kingdom. In 2006 Russia and China were invited to join the organization.

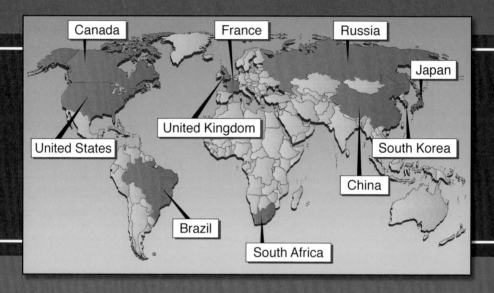

Sources: U.S. Department of Energy, "Gen IV Nuclear Energy Systems," 2008. http://nuclear.energy.gov.

Key People and Advocacy Groups

Roger Billings: In 1965 at the age of 16, Billings converted his father's old Model A Ford to run on hydrogen. In later years he built hydrogen-powered forklifts, tractors, buses, dozens of cars, and a home. Widely known for his work as a promoter of hydrogen energy, Billings was nicknamed the "Hydrogen Man" by *Omni* magazine and "Dr. Hydrogen" by *Time*.

Chris Borroni-Bird: This British-born scientist, designer, and innovator is the director of Advanced Technology Vehicle Concepts for General Motors in Warren, Michigan. As one of the world's leading experts on fuel cells, Borroni-Bird worked on GM's AUTOnomy and Hy-wire concept vehicles. These are the first vehicles designed from the ground up with fuel cell propulsion systems and also the first to allow steering, braking, and other vehicle systems to be controlled electronically rather than mechanically.

Ulf Bossel: With degrees in fluid mechanics and thermodynamics, German-born Bossel has been involved with renewable energy research and development since the 1970s. He began working to improve fuel cells in 1987 and later established himself as a fuel cell consultant for companies in Europe, Japan, and the United States. Today he heads the Bossel European Fuel Cell Forum in Lucerne, Switzerland.

George W. Bush: As president of the United States from 2001 to 2009, Bush was a major promoter of hydrogen energy. In 2003 he directed $1.2 billion in government spending to fund the Hydrogen Fuel Initiative for research into hydrogen production and storage, fuel cell development, and related fields. In 2004 Bush implemented the Nuclear Hydrogen Initiative to demonstrate the commercial-scale production of hydrogen using nuclear power by 2017.

Al Gore: Gore was vice president of the United States from 1993 to 2001. After leaving office, he dedicated his life to drawing public attention to the consequences of global warming. The presentation he gave to students, government officials, and environmentalists was made into the film *An Inconvenient Truth*, which broke box office records for a documentary movie in 2006 and won an Academy Award.

The Green Hydrogen Coalition: The Green Hydrogen Coalition was formed to oppose President George W. Bush's Hydrogen Fuel Initiative and Nuclear Hydrogen Initiative. Made up of eight of the nation's leading environmental, consumer, and public policy organizations, members of the group believe that hydrogen can solve America's energy problems only if it is produced from renewable energy and not from fossil or nuclear fuels.

Amory P. Lovins: Lovins is an author and expert on renewable energy and energy efficiency. As founder, chair, and chief scientist of the Rocky Mountain Institute, Lovins has promoted hydrogen and renewable energy to heads of multinational corporations, government officials in six countries, and leaders at the United Nations. Lovins is also chief designer of the Hypercar, a futuristic vehicle powered by hydrogen fuel cells.

Robert Plarr and Victoria Peters: Plarr is an ex-marine who designed and built a solar- and hydrogen-powered underground home called Angel's Nest in New Mexico. Peters is a former actor and a cofounder, cofinancier, and spokesperson for Angel's Nest. The couple's mission is to promote sustainable living technologies utilizing buildings that capture, purify, and reuse their own clean water and process some of it into hydrogen power.

Jeremy Rifkin: As an economist and the best-selling author of *The Hydrogen Economy*, Rifkin has been a leading voice in the promotion of fuel cells and hydrogen energy. Rifkin is also a fellow at the Wharton School's Executive Education Program, where he lectures business leaders and government officials on trends in science and technology. He pursues similar work as president of the Foundation on Economic Trends in Washington, D.C.

Joseph J. Romm: Romm is an author, lecturer, and scientist whose field of expertise is global warming, greenhouse gases, and energy security. As author of *The Hype About Hydrogen*, Romm is a leading skeptic concerning hydrogen. He believes the money spent on hydrogen research should be directed toward other forms of renewable energy along with high-efficiency hybrid automobiles.

James Woolsey: Woolsey is a former head of the Central Intelligence Agency who now chairs the Clean Fuels Foundation, a group that promotes hybrid vehicles. He believes it is in the national security interests of the United States to reduce global warming and wean the nation from imported oil. Woolsey does not believe hydrogen should have any role in America's energy future because it requires too much electricity to produce and store.

Robert Zubrin: Zubrin is an aerospace engineer, author, and advocate of manned Mars exploration. With degrees in mathematics, aeronautics, astronautics, and nuclear engineering, Zubrin is a hydrogen critic who believes America needs to abandon what he calls the false promise of the hydrogen age. He believes hydrogen is too energy intensive to produce, difficult to store, and inefficient for vehicle use.

Chronology

1963
Australian electrochemist John O'M. Bockris coins the term "hydrogen economy" to describe a system of harmonious, balanced, interacting elements that would provide hydrogen power to the industrial world.

1766
British scientist Henry Cavendish recognizes the energy content of pure hydrogen and labels it "inflammable air."

1923
British scientist J.B.S. Haldane is the first to propose using windmills to electrolyze water into hydrogen and oxygen.

1750   1800   1850   1900   1950

1965
American astronauts spend a record eight days in space in a command module powered by a hydrogen fuel cell.

1820
English inventor William Cecil presents a paper to the Cambridge Philosophical Society titled "On the Application of Hydrogen Gas to Produce Moving Power in Machinery."

1979
On March 29 it is discovered that the core of a nuclear reactor at the Three Mile Island power plant in Harrisburg, Pennsylvania, is in the process of a partial meltdown.

2006
Bryan Beaulieu builds a 6,000-square-foot (557 sq. m), $2 million solar- and hydrogen-powered house in Scottsdale, Arizona, that is completely energy self-sufficient.

1986
In April the worst disaster in nuclear history takes place when there is a meltdown at the Chernobyl power plant in Ukraine.

2008
More than 200 hydrogen-powered vehicles are on roads in the United States, with about 175 in California.

2004
German automaker BMW wins nine world speed records for cars with hydrogen-powered internal combustion engines.

1990   1995   2000   2005   2010

2003
President George W. Bush implements the Hydrogen Fuel Initiative, which dedicates $1.2 billion over five years to research hydrogen production, storage, and fuel cell technologies.

2009
A small number of Southern California drivers lease Honda FCX Clarity fuel cell vehicles for $600 a month.

2007
A Ford Buckeye powered by a hydrogen fuel cell reaches the speed of 207.2 miles per hour (333.5kph), setting a world record for the fastest fuel cell vehicle.

Related Organizations

Alternative Fuels and Advanced Vehicles Data Center

1000 Independence Ave. SW

Washington, DC 20585

phone: (800) 342-5363

fax: (202) 586-4403

Web site: www.afdc.energy.gov

The Alternative Fuels and Advanced Vehicles Data Center is run by the U.S. Department of Energy. It has information on fuel cell vehicles and other alternative fuel vehicles. It gives basic information on non-petroleum fuels, with pages on hydrogen and alternative fuel infra-structure. The Web site has an interactive tool to compare the technical characteristics of various fuels. The center publishes *Clean Cities Now* magazine and various documents and research papers concerning alternative fuels.

American Hydrogen Association (AHA)

2350 W. Shangri La Rd.

Phoenix, AZ 85029

phone: (602) 328-423

e-mail: quest@clean-air.org

Web site: www.clean-air.org

The mission of the AHA is to provide up-to-date information about worldwide developments concerning hydrogen, solar, wind, biomass, en-ergy conversion, and the environment. Its goal is to help establish the re-newable hydrogen energy economy by the year 2010. To achieve this goal the AHA is working with environmental groups, industry, communities, and schools to promote understanding of hydrogen technology while help-ing create a marketplace for pollution-free hydrogen energy. Its bimonthly newsletter, *Hydrogen Today*, features information on hydrogen, solar, and renewable energy.

California Fuel Cell Partnership (CFCP)

21865 Copley Dr., Suite 1137

Diamond Bar, CA 91765

phone: (909) 396-3388

fax: (909) 396-3387

e-mail: info@cafcp.org

Web site: www.drivingthefuture.org

The CFCP is committed to promoting fuel cell vehicles as a means of moving toward a sustainable energy future, increasing energy efficiency, and reducing or eliminating air pollution and greenhouse gas emissions. The CFCP is composed of 33 member organizations that include auto manufacturers, energy providers, government agencies, and fuel cell technology companies. The group publishes press releases and news clips concerning hydrogen power.

Clean Fuels Development Coalition (CFDC)

4641 Montgomery Ave., Suite 350

Bethesda, MD 20814

phone: (301) 718-0077

fax: (301) 718-0606

e-mail: cfdcinc@aol.com

Web site: www.cleanfuelsdc.org

The CFDC is a nonprofit organization that actively supports the increased production and use of fuels that can reduce air pollution and oil imports. The CFDC works to support clean fuel regulations and new technologies that will help the United States meet its complex energy demands. One of the coalition's primary goals is public education aimed at generating support for environmentally safe fuels. CFDC publications are categorized into fact books, issue briefs, white papers, industry and public forums, and general information.

European Hydrogen Association (EHA)

Gulledelle 98

1200 Bruxelles

Belgium

phone: 32 2 763 25 61

fax: 32 2 772 50 44

e-mail: info@h2euro.org

Web site: www.h2euro.org

The EHA was formed to promote hydrogen power in Europe by working with elected officials, industrial companies, teachers, students, research scientists, the media, and the general public. The group publishes press releases, studies, and presentations.

Hydrogen Now!

Engineering Research Center

Colorado State University

Fort Collins, CO 80523

phone: (970) 491-7189 or (866) 464-2669

e-mail: info@hydrogennow.org

Web site: www.HydrogenNow.org

Hydrogen Now! is an international organization dedicated to making the hydrogen economy a reality. The organization is composed of global leaders in industry, government, and the scientific community. The group's Web site provides links to *H2Nation* magazine, the *Alternative Energy News Source, Hydrogen & Fuel Cell Letter*, and other publications.

National Hydrogen Association (NHA)

1211 Connecticut Ave. NW, Suite 600

Washington, DC 20036-2701

phone: (202) 223-5547

fax: (202) 223-5537

e-mail: info@hydrogenassociation.org

Web site: www.hydrogenassociation.org

The NHA was formed to lead the transition from a fossil fuel–based energy infrastructure to one that is hydrogen-based. The NHA has over 100 members, including representatives from the automobile industry; the fuel cell industry; aerospace; federal, state, and local governments; and energy providers. The group's mission is to foster the development of hydrogen technologies and their utilization in industrial, commercial, and consumer applications and promote the role of hydrogen in the energy field. It publishes fact sheets, hydrogen headlines, and the Clean Energy Events calendar.

The National Renewable Energy Laboratory (NREL)

1617 Cole Blvd.

Golden, CO 80401-3393

phone: (303) 275-3000

Web site: www.nrel.gov

The NREL is the U.S. Department of Energy's laboratory for renewable energy research, development, and deployment, and a leading laboratory for energy efficiency. The laboratory's mission is to develop renewable energy and energy-efficiency technologies, advance related science and engineering, and transfer knowledge and innovations to address the nation's energy and environmental goals. The lab's Hydrogen and Fuel Cell Research Department deals exclusively with issues related to hydrogen power. The organization publishes dozens of comprehensive research papers concerning these technologies, many of them available for free online.

Renewable Energy Policy Project (REPP)

1612 K St. NW, Suite 202

Washington, DC 20006

phone: (202) 293-2898

fax: (202) 298-5857

e-mail: info2@repp.org

Web site: www.repp.org

The REPP provides information about hydrogen, solar, biomass, wind, hydro, and other forms of green energy. The goal of the group is to accelerate the use of renewable energy by providing credible facts, policy analysis, and innovative strategies concerning renewables. The project has a comprehensive online library of publications dedicated to these issues.

U.S. Department of Energy
Hydrogen, Fuel Cells and Infrastructure Technologies

1000 Independence Ave. SW

Washington, DC 20585

phone: (800) 342-5363

fax: (202) 586-4403

Web site: www1.eere.energy.gov/hydrogenandfuelcells

The Department of Energy's overarching mission is to advance the national, economic, and energy security of the United States; to promote scientific and technological innovation in support of that mission; and to ensure the environmental cleanup of the national nuclear weapons complex. The department's Hydrogen, Fuel Cells and Infrastructure Technologies program deals exclusively with hydrogen production, delivery, and storage; fuel cells; and other related issues. The department publishes three newsletters and dozens of comprehensive research papers.

For Further Research

Books

Jack Erjavec, *Hybrid, Electric, and Fuel-Cell Vehicles.* Clifton Park, NJ: Thomson Delmar Learning, 2007.

Rex A. Ewing, *Hydrogen—Hot Stuff, Cool Science: Discover the Future of Energy.* Masonville, CO: PixyJack, 2007.

Gavin D.J. Harper, *Fuel Cell Projects for the Evil Genius.* New York: McGraw-Hill/TAB Electronics, 2008.

Geoffrey B. Holland and James J. Provenzano, *The Hydrogen Age.* Layton, UT: Gibbs Smith, 2007.

Krishnan Rajeshwar, Robert McConnell, and Stuart Licht, *Solar Hydrogen Generation: Toward a Renewable Energy Future.* New York: Springer, 2007.

Jeremy Rifkin, *The Hydrogen Economy.* New York: Jeremy P. Tarcher/Putnam, 2002.

Joseph J. Romm, *The Hype About Hydrogen.* Washington, DC: Island, 2005.

Vijay V. Vaitheeswaran, *Power to the People.* New York: Farrar, Straus, and Giroux, 2003.

Periodicals

Janice Arenofsky, "The Hydrogen House: Fueling a Dream," *E*, January/February 2006.

Melissa Block et al., "GM Official: Hydrogen-Powered Cars Impractical" (transcript), *All Things Considered*, March 5, 2008.

D. Castelvecchi, "Platinumfree Fuel Cell," *Science News*, October 20, 2007.

Richard Day, "Emission-Free Europe: Hydrogen Projects, from Iceland to Italy," *E*, January/February 2007.

James R. Healey, "Fuel-Cell Cars Great, but You Can't Have One," *USA Today*, November 9, 2007.

Jay Leno, "Will Hydrogen Fuel Our Future? Jay Spends 10 Days Behind the Wheel of BMW's Hydrogen 7," *Popular Mechanics*, January 2008.

Steve Maich, "Reinventing Your Wheels: The Non-Polluting Hydrogen Car Has Finally Arrived . . . Almost," *Maclean's*, March 5, 2007.

James McCraw, "Fuel Cell Land Speed Record Setter," *Popular Mechanics*, December 2007.

Kara Rowland, "Fuel-Cell Vehicles Stalled by Price Tag," *Washington Times*, February 20, 2008.

Mark K. Solheim, "Who Needs Gas Engines?" *Kiplinger's Personal Finance*, October 2006.

Luther Turmelle, "Hydrogen-Powered Autos in Spotlight at Area Tour," *New Haven (CT) Register*, August 13, 2008.

Lawrence Ulrich, "Drive the Future, No Money Down," *Popular Science*, September 2007.

Tara Weingart, "Road Test: Honda FCX Concept; Step on the, er, Gas," *Newsweek*, December 18, 2006.

P. Weiss, "Hydrogen Hopes in Carbon Shells," *Science News*, August 19, 2006.

Chris Woodyard, "Tokyo Show Proves Hydrogen Is Popular," *USA Today*, October 24, 2005.

Internet Sources

Angel's Nest, "Angel's Nest—Renewable Energy and Sustainable Architecture." www.angels-nest.org.

AutoblogGreen, Sebastian Blanco, ed., www.autobloggreen.com.

Hydro Kevin, "Hydrogen Cars & Vehicles," Hydrogen Cars Now. www.hydrogencarsnow.com, 2008.

Hydrogen.gov, "The President's Hydrogen Fuel Initiative." www.hydrogen.gov/thepresidentshydrogen_fi.html.

Rocky Mountain Institute, "What Is a Hypercar Vehicle?" www.rmi.org/sitepages/pid191.php, 2008.

Source Notes

Overview

1. Quoted in Doug Main, "Scientists Seek to Solve Hydrogen Storage Problems," *Washington University Record,* December 2, 2005. http://record.wustl.edu.
2. Quoted in Main, "Scientists Seek to Solve Hydrogen Storage Problems."
3. Quoted in Jeremy Rifkin, *The Hydrogen Economy.* New York: Jeremy P. Tarcher/Putnam, 2002, p. 207.
4. Quoted in Goska Romanowicz, "London's Trial Hydrogen Buses Multiply," EDIE. www.edie.net.
5. Quoted in Robert S. Boyd, "Hydrogen Cars May Be a Long Time Coming," McClatchy Washington Bureau, May 15, 2007. www.mcclatchydc.com.
6. Geoffrey B. Holland and James J. Provenzano, *The Hydrogen Age.* Layton, UT: Gibbs Smith, 2007, p. 129.
7. Holland and Provenzano, *The Hydrogen Age,* p. 138.
8. George W. Bush, "Hydrogen Fuel Initiative Can Make 'Fundamental Difference,'" The White House, February 2003. www.whitehouse.gov.
9. Holland and Provenzano, *The Hydrogen Age,* pp. 138–39.
10. Quoted in Hydro Kevin, "Cost of Hydrogen Fueling Infrastructure," Hydro Cars & Vehicles. www.hydrogencarsnow.com.

Can Hydrogen Vehicles Reduce Dependence on Fossil Fuels?

11. Quoted in James McCraw, "Fuel Cell Land Speed Record Setter," *Popular Mechanics,* December 2007, p. 114.
12. Holland and Provenzano, *The Hydrogen Age,* p. 210.
13. Taiyo Kawai et al., *The Hydrogen Energy Transition.* Burlington, MA: Elsevier, 2004, p. 67.
14. Quoted in Kara Rowland, "Fuel-Cell Vehicles Stalled by Price Tag," *Washington Times,* February 20, 2008, p. A-1.
15. Quoted in Holland and Provenzano, *The Hydrogen Age,* p. 211.
16. Quoted in Holland and Provenzano, *The Hydrogen Age,* p. 13.
17. Quoted in Auto Industry, "Clean Fuels Foundation Chairman Discounts Hydrogen's Potential," November 29, 2007. www.autoindustry.co.uk.
18. Quoted in Rifkin, *The Hydrogen Economy,* p. 209.
19. Lucas Graves and Larry Burns, "Why GM Is High on Hydrogen," *Wired,* July 2005. www.wired.com.
20. Quoted in Robert Lester, "Analysis: Is Hydrogen a Viable Fuel Alternative Despite Recent Criticism?" *Marketing Week,* December 13, 2007, p. 8.

Can Hydrogen Supply the World's Electricity Needs?

21. Quoted in Peter Hoffmann, *The Forever Fuel: The Story of Hydrogen.* Boulder, CO: Westview, 1981, p. 83.
22. Quoted in Rifkin, *The Hydrogen Economy,* p. 172.
23. Quoted in Maggie Galehouse, "One Man's Castle Runs on Hydrogen," MSNBC, May 30, 2005. www.msnbc.msn.com.
24. Quoted in Diana Heil, "Harnessing Hydrogen to Fuel the Future," *New Mexican,* August 24, 2005. www.fuelcellsworks.com.

25. Quoted in Nathan Diebenow, "Home Sweet, Homeland Security," *Lone Star Iconoclast*, October 2, 2006. http://lonestaricon.com.

26. Robert Plarr, "Angel's Nest—Renewable Energy and Sustainable Architecture," Angel's Nest. www.angels-nest.org.

27. Quoted in Diebenow, "Home Sweet, Homeland Security."

28. Ulf Bossel, "The Hydrogen Illusion," European Fuel Cell Forum, March/April 2004. www.efcf.com.

How Will Hydrogen Use Impact Global Warming?

29. Nobelprize.org, "The Nobel Peace Prize 2007." http://nobelprize.org.

30. Quoted in Kristin Arna Bragadottir, "Iceland's Hydrogen Ship Heralds Fossil-Free Future," Reuters, January 23, 2008. www.reuters.com.

31. Quoted in Greenpeace USA, "New Greenpeace Report Exposes CCS as a Dangerous Distraction," May 5, 2008. www.greenpeace.org.

32. Rex A. Ewing, *Hydrogen—Hot Stuff, Cool Science: Discover the Future of Energy.* Masonville, CO: PixyJack, 2007, p. 155.

What Is the Future of Hydrogen Power?

33. Holland and Provenzano, *The Hydrogen Age*, p. 322.

34. Quoted in Sam Carana, "Breakthroughs Open Door to Hydrogen Economy," Gather, August 1, 2008. www.gather.com.

35. U.S. Department of Energy, "Nuclear Hydrogen Initiative." www.ne.doe.gov.

36. Quoted in American Society of Plant Biologists, "Members Melis and Ghirardi Find Switch for Hydrogen Production." www.aspb.org.

37. Jeremy Rifkin, "How Hydrogen Empowers Fair Globalization," *Globalist*, January 11, 2003. www.theglobalist.com.

38. Quoted in Rifkin, *The Hydrogen Economy*, p. 177.

List of Illustrations

Index

About the Author

Stuart A. Kallen is a prolific author who has written more than 250 non-fiction books for children and young adults over the past 20 years. His books have covered countless aspects of human history, culture, and science from the building of the pyramids to the music of the twenty-first century. Some of his recent titles include *How Should the World Respond to Global Warming, Romantic Art*, and *World Energy Crisis.* Kallen is also an accomplished singer-songwriter and guitarist in San Diego, California.